Laboratory Activities Manual for Human Biology

Craig Clifford

Northeastern State University

SAUNDERS COLLEGE PUBLISHING
HARCOURT BRACE COLLEGE PUBLISHERS

Fort Worth   Philadelphia   San Diego   New York   Orlando
San Antonio   Toronto   Montreal   London   Sydney   Tokyo

Copyright © 1998 by Saunders College Publishing

All rights reserved. No part of this publication may be reproduced or transmitted in any form or by any means, electronic or mechanical, including photocopy, recording, or any information storage and retrieval system, without permission in writing from the publisher.

Requests for permission to make copies of any part of the work should be mailed to Permissions Department, Harcourt Brace & Company, 6277 Sea Harbor Drive, Orlando, Florida 32887-6777.

Vice President/Publisher: Emily Barrosse
Acquisitions Editor: Edith Beard Brady
Product Manager: Erik Fahlgren
Developmental Editor: Lee Marcott
Production Manager/Director of EDP: Arnold Lynch Jr.
Art Director, Text Designer, and Cover Designer: Kathleen Flanagan
Photo Researcher: Jane Sanders

Cover Image Credit: The Dance II, by Henri Matisse, © 1997 Succession H. Matisse, Paris/Artists Rights Society (ARS), New York; additional rights from the Hermitage Museum, St. Petersburg, Russia/Giraudo, Paris/SuperStock.

Printed in the United States of America

Lab Manual to Accompany Human Biology

0-03-025191-5

Library of Congress Cataloging-in-Publication Data
was not available at the time of printing

7890123456 202 10987654321

Dedication

This effort is dedicated to
supportive family members and friends,
especially
my father who kept asking,
"Is it done yet?"

Table of Contents

Preface

Section 1 Introduction to the Study of the Human Body	1
Anatomy and Physiology	1
Anatomy First, Then Physiology	2
Anatomical Position	2
Body Planes	4
Directional Terms	4
Use of Directional Terms	5
Homeostasis	6
Negative Feedback	8
Journal Activity	9
Handling Body Fluids	10
Section 2 Cells and Tissues	13
Use of the Microscope	13
Slide Preparation	15
Cytology: The Study of Cells	16
An Atlas of Tissue Types	19
Diffusion and Osmosis	22
Observing Diffusion and Osmosis	24
Cellular Division	28
The Cell Cycle	29
Section 3 The Integument	33
Structure and Histology of the Skin	33
Two-Point Discrimination	37
Hot, Cold, and Touch Receptors	39
Triple Response	41
Rate of Nail Growth	43
Demonstration of the Sweat Glands	44
Section 4 Bone and the Skeletal System	47
The Histology of Compact Bone	48
The Skeleton	50
The Skull	53
The Cranial Bones	55
The Facial Bones	56
The Rib Cage and Vertebral Column	57
The Upper Limb (The Arm)	59
The Lower Limb (The Leg)	62

Section 5 The Muscular System	67
The Histology of Muscle	67
The Physiology of Muscle Contraction	70
Muscle Contraction: A Play	72
The Muscles of the Body	74
Naming the Muscles	74
Learning the Superficial Muscles	75
Electromyography	84
Section 6 The Endorine System	87
Types of Hormones	88
Control of Hormone Release	89
The Pituitary Gland	89
The Thyroid Gland	90
The Parathyroid Glands	92
The Adrenal Glands	92
The Pancreas	95
Determination of Blood Glucose Concentration	95
The Gonads	97
Section 7 The Nervous System—Part 1	99
Architecture of the Nervous System	99
Anatomy of the Spinal Cord	100
The Meninges	104
Physiology of the Spinal Cord: Reflexes	104
Reflex Demonstration	106
Patellar or Knee-Jerk Reflex	106
Ankle-Jerk or Achilles Reflex	108
Plantar and Babinski Reflexes	109
Anatomy of the Brain	111
The Brain Stem	111
The Reticular Formation	112
The Cerebellar Peduncles	113
The Diencephalon	113
The Cerebrum	113
The Cerebellum	114
The Cranial Nerves	115
Section 8 The Nervous System—Part 2	117
Sensory Physiology	117
The Gustatory Sense	117
Demonstrating the Gustatory Sense	119
The Olfactory Sense	121
Demonstrating the Olfactory Sense	121
The Visual Sense	124
Structure of the Eyes	124
Determination of Visual Acuity	125
Blind Spot Determination	128
Color Blindness	129
The Auditory Sense	130
The Hearing Test	132
Hearing Deficit and Diagnosis	133

Section 9 The Circulatory System — 137

- Architecture of the Circulatory System — 137
- Demonstrating the Valves in the Veins — 142
- Measuring Blood Pressure — 144
- Determining Blood Pressure — 145
- Determining Pulse and Pulse Pressure — 149
- Anatomy of the Heart and Vasculature — 151
- Listening to the Heart — 152
- Control of the Heartbeat — 154
- Suggested Activity: ECG — 156
- Blood and Blood Parameters — 156

Section 10 The Digestive System — 161

- Anatomy of the Digestive Tract — 163
- Enzymes — 169
- Inorganic Catalysis versus Enzymatic Catalysis — 170
- Enzymatic Activity — 171
- Nutritional Analysis — 174
- Nutritional Needs — 175
- Methods for Determining Body Composition — 178

Section 11 The Respiratory System — 185

- Architecture of the Respiratory System — 185
- The Nose — 185
- The Pharynx and Larynx — 187
- The Trachea and Bronchial Tree — 188
- Respiratory Muscles and Breathing Control — 189
- Breath Holding — 190
- Respiratory Volumes — 191
- Spirometry to Determine Respiratory Volumes — 192
- Determining the Metabolic Rate — 195
- Practical Considerations in Metabolic Measurements — 195
- Oxygen-Carrying Capacity of Blood — 198
- Total Red-Blood-Cell Count — 199
- Hemoglobin Concentration — 199
- Suggested Activity: Complete Blood Count/Lab Work-Up — 199

Section 12 The Urinary System — 201

- The Kidneys — 199
- Structure of the Nephron — 204
- Urine Formation — 206
- Analysis of Urine — 209

Section 13 The Reproductive System — 213

- Meiosis: Reduction Division — 213
- The Male Reproductive System — 216
- Stages of Spermatogenesis — 217
- Glands of the Male Reproductive System — 220
- External Male Genitalia — 221
- Erection and Ejaculation — 222
- Anatomy of a Sperm Cell — 223
- The Female Reproductive System — 224
- Stages of Oogenesis — 225
- Birth Control Devices — 227

Mammary Glands	230
Pregnancy	231
Main Features of Embryonic and Fetal Development	231

Section 14 Genetics 233

Basic Principles of Genetics	233
How do Genes Work?	235
Testcrosses and Pedigrees	239

Appendix Environmental Concerns 243

Preface

This lab activities manual is intended to be a supplement to your lecture and its text. It also is unlikely that you would be using this lab manual without taking a human biology course at the same time. Therefore, plan to use the information from lectures and the lecture text to help you in lab.

Anatomy, physiology, genetics, environmental issues and other topics covered in your lecture text are important in understanding human biology. However, even the most dynamic lecturer presenting information via multimedia formats is still conveying lots of information. It is difficult to give hands-on attention in lecture. That's where the lab activities manual comes in. In lab with models, charts, observations, measurements, and experimental activities, the information conveyed in lecture should come alive.

Your body is a dynamic, complicated machine, and it doesn't come with a manual. This lab activities manual will provide lots of anatomical and physiological details that will help you understand your structure and function. The understanding of the function of the parts of a machine is necessary to trying to fix problems that arise. An understanding of the workings of the human body is necessary to practicing any of the areas of medicine.

Format of This Manual

This manual is not organized like most manuals. It is entitled a lab activities manual because it contains many short, easy-to-follow exercises that help the student understand and appreciate anatomy, physiology, genetics, and environmental issues. Each exercise usually has questions throughout and at the end of each procedure. The purpose of the questions is to stimulate you to think about the information being presented or discovered in the activity. The answers to some questions are obvious from their preceding discussion. The questions also may be a means to guide you into the next phase of the discovery process. Some questions may even be open-ended to stimulate speculation about subjects that are not fully explained in the text. **Do not be frustrated by the questions; use them as tools to better understand these topics.**

Learning Styles/Learning Strategies

There is much educational research into the different ways in which people learn. We have all heard about visual versus auditory learners. Some people apparently pick up and retain new material just by hearing the material being presented, as in a lecture format. Others require that something concrete be before them so they can see the material about which information is being presented.

Obviously, books in general focus on the visual learning styles. Regardless of the way you might learn best, there is a certain amount of repetition required to take in and retain large amounts of detailed material, as is required in subjects like those covered in this course. As many of the terms you read may be unfamiliar to you, write the words several times in the journal that is suggested in Section One, in a notebook or even on scrap paper. The act of writing the term helps implant it in your memory. Writing is a more active/interactive process than is simply reading.

Because you will be learning a lot of new terms in this course, flash cards may be helpful. Obtain or make small uniform size cards. On one side, write the word or term that you are trying to learn. On the other side of the card, write the definition or explanation of the term. Read the term, read the definition. Read the term again, and then try to repeat the definition out loud if possible, or write it down. Keep the cards handy so that if there is some unexpected study time in your day, you can take advantage of it by reviewing the material on the cards.

We all develop bad habits from time to time. One bad habit may be becoming too comfortable with our surroundings. If there are too many convenient distractions in the place where you typically study, find a new location. Find an empty classroom or study hall in a school facility. Use the chalkboard as a convenient way to write the material being studied. Review your notes and then write the key points on the board. Try to explain the material to a fellow student from the class.

Very few people can stick to a single activity, especially if mentally demanding, for a long period of time. Don't try to study for hours on end. Decide for yourself a reasonable length of time for study, and then take short breaks after studying for a set time. Get up, move around, and stretch. Get a drink of water and then return to the study area refreshed for another round at the books.

Acknowledgment

This manual would not have been possible without the help of the many students that I have interacted with over the years. Thanks to those who let me "experiment" with them. A special thanks is given to John Moles and Brad Brown for posing for photographs. Many photos were provided by Northeastern State University, Photographic Services – Mike Brown. Thanks also to the dean and staff of the NSU College of Science, Mathematics and Nursing for general office services and support.

Final Note

I hope that whatever you experience using this manual helps you to better understand your body and the impact of humans on the world. I also hope that you enjoy this course and have fun as you learn.

Craig Clifford
April 1998

Section 1

Introduction to the Study of the Human Body

Note to the Student

You are beginning a course in human biology. Many people shy away from the sciences and feel they cannot handle the material. You probably are not interested in a health-related career that specifically requires more advanced coverage of this material. This will be an introductory course that may help you to decide if you are truly interested in a more in-depth study of this field. Most students in this course will take it as one of very few in the sciences during their college career. I hope you learn much and will leave the course with a general and comfortable knowledge of your body.

Anatomy and Physiology

Both your text and this lab manual have "human biology" in the name, but they deal essentially with human anatomy and physiology. Therefore, we start with an introductory review of anatomy and physiology as a prelude to our discussion of the human body.

Anatomy (*ana* (G), up; *tomy* (G), cut) refers to the structure of an organism, whether it is an animal or a plant. The term also refers to the study or science of the structure of an organism and its parts. Because scientists are very precise in their terminology, you will be required to learn many terms and their meanings. Do not be discouraged. Through a systematic approach using medical terminology and attention to detail, you will soon be able to talk the language of anatomy.

Physiology (*physio* (G), nature; *logy* (G), study of) is the other area we will discuss. The term refers to the science dealing with the essential life processes, activities, and functions of an organism. The key word in this description is *function*. Physiology addresses how an organism carries out the tasks that maintain life.

The anatomy and physiology of any organism can be studied. We focus on the anatomy and physiology of the human (i.e., human biology). There are very good reasons to study other groups of organisms, but as humans, we are especially concerned with our own anatomy and physiology.

QUESTION: The anatomy and physiology of any living organism can be studied, but what is the difference between anatomy and physiology as subject areas? In other words, how might our approach to the study of anatomy differ fundamentally from our approach to the study of physiology? (Hint: Every group of organisms has its own special anatomy, whereas physiology is the same for all the basic systems.)

Observations:

Conclusion:

Anatomy First, Then Physiology

There is no absolute prescription regarding the order in which anatomy and physiology must be studied. Anatomy typically is discussed first, however; then physiology can be studied and related to the anatomy. Not all courses in physiology require an anatomy course as a prerequisite, but it is assumed that when students study physiology, they already have a basic knowledge regarding the anatomy of the system under consideration. Otherwise, it would be like trying to understand the workings of an automobile with no concept of how an engine is constructed. The explanation would refer to parts and processes that would overwhelm students without conveying what makes the automobile move. Thus, as we proceed from system to system, each chapter will have sections on anatomy followed by sections on physiology.

Anatomical Position

Most fields of study have certain underlying principles on which the subject is based and our understanding depends. Though it greatly oversimplifies the studies of anatomy and physiology, we can name a few, very important principles that must be appreciated before working in either area. In anatomy, a very important principle is that of anatomical position; in physiology, an important concept is homeostasis.

Anatomy is the study of the structure of the parts of the human body. To discuss structures, not only the feature but also its location must be described. There needs to be some benchmark or reference point so that we can consistently say where a structure is located. This point of reference is the **anatomical position** (Figure 1.1).

> **Principle:** Anatomical position is defined as the body standing erect, with the feet apart and the toes facing forward. Additionally, the arms are at the side, with the palms and the head facing forward.

Figure 1.1
Anatomical Position.
This person is standing in anatomical position. The body is erect, with arms at the side, palms facing forward, feet spaced apart, and toes facing forward. It is useful to picture the body in this position, because reference to body parts then will be in relation to this orientation.

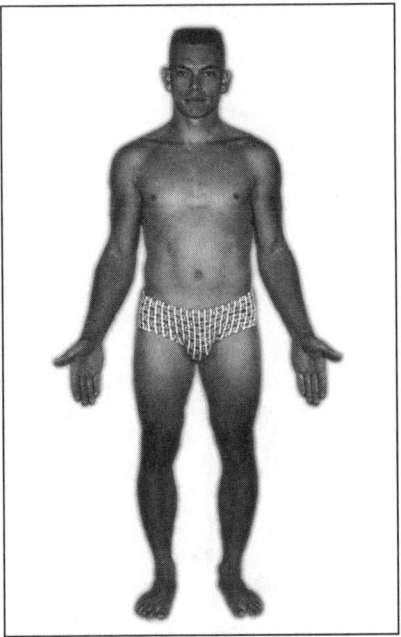

It can be assumed that when the position of any body part is being described, it is in reference to the head, torso, arms, and legs in the anatomical position. This way, everyone will understand the position of structures, because their location will be consistently described.

Sections, planes, and directional terms also are important to anatomy, because with these terms, the location of a structure can be described. Often, it is useful to imagine cutting the body along a line or a plane to appreciate the location of one body part in relation to another. Such a cut is a **section.** Note that the sectioning described in this course is an imaginary process. In advanced courses using cadavers, however, students may actually section a body to see the relationship of the enclosed organs, bones, and muscles. With new techniques for preserving and embedding in plastics, many medical schools now have human sections that allow students to see what is inside of the body (Figure 1.2). Modern imaging techniques enable us to "see" inside of the body without cutting at all. Magnetic resonance imaging, for example, reveals impressive detail about our internal anatomy.

Figure 1.2
Embedded Transverse Section of the Human Body.
There are various techniques for using real human remains in anatomy. One new method is to embed an actual transverse or horizontal section of a preserved cadaver in plastic resin, as shown.

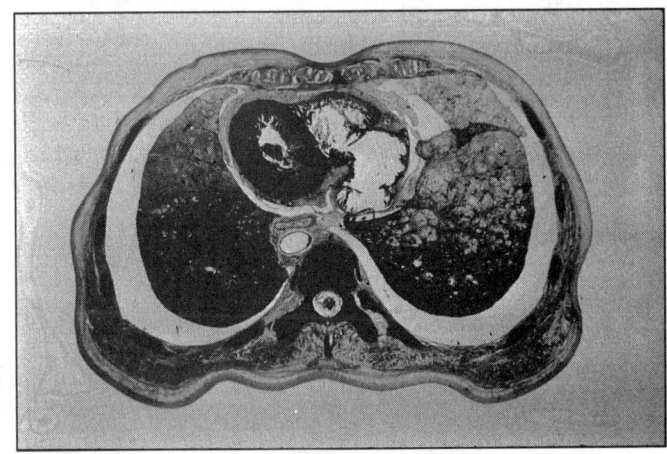

Figure 1.3
Anatomical Sectioning of the Body
Shown are the typical sections used to schematically or actually section the human body for study and to describe the orientation or position of organs and body parts.

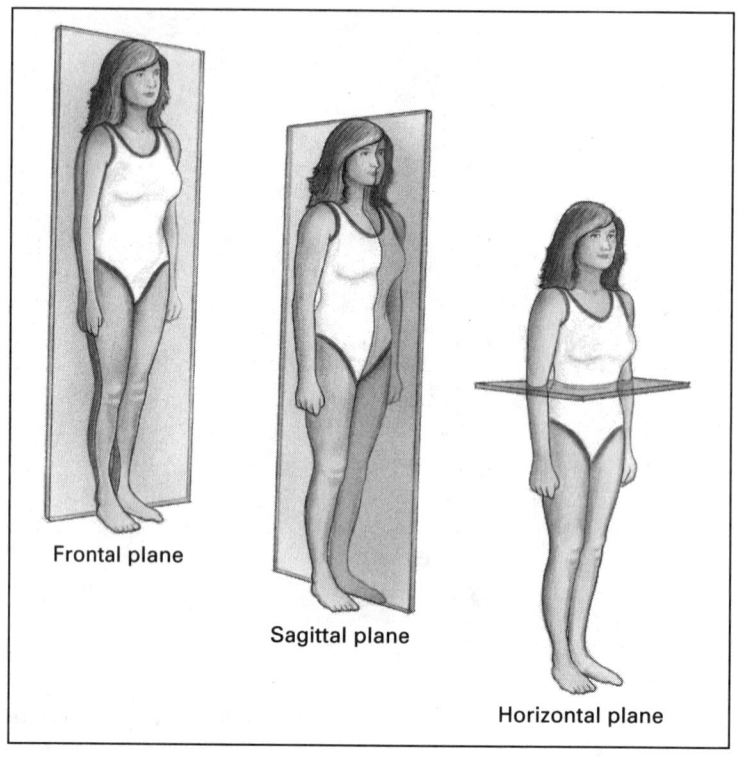

Body Planes

When a part of the body is sectioned, it is done so along planes. Typically, a **plane** is described as an imaginary surface. As described, however, the body can actually be cut along these surfaces, so they are not always imaginary. Typical planes that can be used when sectioning the body or parts of the body include (Figure 1.3):

- **Midsagittal plane:** A vertical plane on the midline that divides the body into equal left and right parts (also known as a median plane).
- **Sagittal plane:** A vertical plane that is not along the midline, resulting in unequal left and right parts.
- **Frontal plane:** A vertical plane that divides the body into unequal front and back portions (also known as a coronal plane).
- **Transverse plane:** A horizontal plane that cuts the body into two portions somewhere between the head and the feet; similar to most of the other planes, these parts are unequal (also known as a cross-section or x-section).

A thorough understanding of the different planes will facilitate forming a mental image of the location of the various body parts as you read descriptions in the text or lab manual.

Directional Terms

Directional terms allow us to describe the areas of the body and the location of parts. The following terms are used most commonly, and they will allow you to discuss the location of body parts. When using these terms, you will be comparing the location of one structure in relationship to another.

1. **Anterior** (*antero* (L), in front of).

2. **Posterior** (*post* (L), behind).

Note: This pair of terms refers to the front portion of the body (i.e., the *anterior*) and the rear portion (i.e., the *posterior*). They also can be used to describe a structure that is in front of (i.e., anterior to) or behind (i.e., posterior to) another.

3. **Medial** (*media* (L), the middle).
4. **Lateral** (*lateral* (L), the side).

Note: These two terms refer to the side or middle of the body. A structure closer to the midline than another is *medial* to the other. Some feature that is further from the midline than another is *lateral* to the first structure.

5. **Superior** (*super* (L), above).
6. **Inferior** (*infero* (L), low, underneath).

Note: *Superior* means toward the upper parts of the body; *inferior* means toward the lower parts of the body. Superior does not imply superiority or greatness, but does tell us that a structure or process is localized in the upper parts of the body.

7. **Proximal** (*proxim* (L), nearest).
8. **Distal** (*dista* (L), distant).

Note: These two terms are most useful regarding the arms and legs. They refer to where in the limb a structure is located. The part closer to where the limb is attached to the body is the *proximal* structure, whereas a feature that is further away from the limb attachment is *distal*.

Use of Directional Terms

The best way to understand directional terms and how to apply them is to use them when comparing structures.

Directional Terms

Objectives:
- To become familiar with directional terms that refer to different parts of the body.
- To learn how to use directional terms in reference to different parts of the body.

Materials:
- None.

Procedure:

1. Use directional terms to indicate the locations of the following areas or structures:
   - The ears are _____ to the eyes.
   - The head is _____ to the chest.
   - The feet are _____ to the ankles.
   - The hips are _____ to the chest.
   - The knee is _____ to the shin.
   - The navel is _____ to the back.
   - The elbow is _____ to the shoulder.

2. Could any of these areas or structures be described by more than one set of terms?

QUESTION: Not all animals have the same orientation to the ground as humans. Humans walk on only two legs; most other mammals walk on all four legs. Using what you know about other mammals, can directional terms be applied in the same way as in humans?

Observations:

Conclusion:

Homeostasis

Homeostasis is the underlying principle in the field of physiology. When broken into component parts, the word can be translated as "to stay the same" (*homeo* (G), like; *stasis* (G), standing). This might imply that whatever it refers to never changes. The term, however, refers to the fact that even though the internal conditions of the body vary within norms, they still remain basically the same. This concept was described in the nineteenth century by the physiologist Claude Bernard. His term for this phenomenon was ***milieu interieux*** ("internal environment"). Bernard conceptualized the idea that despite fluctuations, the interior of an organism stays about the same (within some normal range of variation). Another physiologist coined the term *homeostasis*. This concept now is described as a state of "dynamic constancy." Though putting *dynamic* and *constancy* together seems to be a contradiction in terms, it accurately describes the limited variation that occurs within our bodies.

> **Principle:** Homeostasis is defined as "a state of physiological equilibrium produced by a balance of functions and chemical composition within an organism." (*The American Heritage Dictionary*, Second College Edition, Houghton Mifflin Co., Boston 1982)

QUESTION: When a person has a fever, we think that person may be sick. In other words, whenever some bodily parameter is "not normal," we suspect an infection or disease in that person. Could we study physiology or could a doctor practice medicine without some predictability in the body's characteristics?

Observations:

Conclusion:

If all bodily processes are kept within some limited range of normality, this implies there are mechanisms that regulate these processes. The generally applied term for these regulatory pathways is **feedback mechanisms,** which implies that some information about the current status of a process or level of a chemical is required before any decision to modify that status or level is made. A feedback mechanism is made up of: (1) a sensor that monitors and therefore detects the current status of the system being controlled, (2) a controller that takes that information and makes a decision as to the appropriate response and, (3) an effector that carries out the decision of the controller (Figure 1.4). Thus, for any parameter in the body, there probably is a mechanism that controls or regulates that parameter. This indicates that decisions made by the controller reverse the change that may be occurring in the parameter; this type of feedback mechanism is said to be **negative.** In other words, the correction performed by the effector, as initiated by the controller, will reverse the change noted by the sensor. Negative does not imply the change is always in a negative direction, but it does imply the change is opposite to the original deviation.

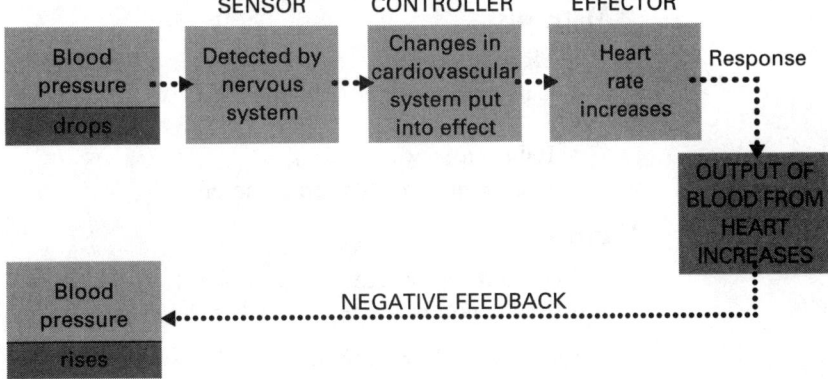

Figure 1.4
Feedback Mechanism.
Shown are the component parts of a feedback mechanism. To illustrate, an example mechanism for control of blood pressure is used. This consists of (1) a sensor to detect the change in the system, and (2) a controller to monitor information from the sensor and then send a signal to (3) an effector to carry out a response that corrects the change detected by the sensor. Through such mechanisms, homeostasis is maintained. Note that the sensor monitors the actions of the effector to complete the circuit.

Negative Feedback

Many phenomena in our natural world and our body demonstrate the principles of homeostasis as controlled by negative feedback. A classic example is the household thermostat. Whether it is winter or summer, either the heater or the cooler will be fighting the environmental temperature outside to maintain a comfortable temperature inside the home. The heater comes on in response to the thermostat detecting that the temperature has dropped below the desired value. This desired value or **setpoint** is that value the system is striving to maintain. The thermostat typically cannot detect minute variations from the setpoint. Usually, the temperature must vary a few degrees from the setpoint before the system can respond; this amount is called the **sensitivity**. The more sensitive the system, the smaller the change that must occur before there is a response. In a similar vein, the heater will stay on longer than is necessary to reach the setpoint, so the temperature will go a bit above the setpoint. The term **range** refers to the total amount the temperature varies from the highest temperature attained to the lowest temperature "allowed."

A laboratory water bath is a device that can maintain the temperature of the contained water. After the heater turns on, the water temperature increases; after the heater turns off, the temperature drops. There is a difficulty in monitoring this process. The temperature variation in a typical laboratory water bath is so small that the familiar $-10°$ to $110°C$ lab thermometer is not capable of showing the variation. The average kitchen fryer/cooker with a variable heat setting, however, is not made to the same standards as the laboratory water bath. Because its temperature variation is much greater, it can be detected with a lab thermometer.

Illustrating Negative Feedback

Objectives:

- To understand homeostasis as the basic premise of physiology.
- To appreciate the role of negative feedback mechanisms in maintaining homeostasis.
- To understand that many everyday processes also function with negative feedback control to maintain a constant state.

Materials:

- Household fryer/cooker with variable heat setting.
- Lab thermometer, $-10°$ to $110°C$ or digital electronic thermometer.

Or

- Laboratory water bath.
- Digital electronic thermometer.

Procedure:

1. Place a thermometer in the cooker (water bath) with the heat level at a low setting. Initially, you will not know what the setting of the water bath is.

2. Observe the temperature variation over three or four cycles of the heater turning on and off. Record the high and low temperatures as indicated by the thermometer when the heater cycles on and off. Depending on the brand of cooker (water bath) used, the heater cycle may be indicated by a small light or you may hear the sound of water boiling over the heating element as it turns on and goes off.

3. Determine the setpoint, sensitivity, and the range of the cooker (water bath).

    Setpoint: _____

    Sensitivity: _____

    Range: _____

QUESTION: Two examples of homeostatic systems already have been mentioned: the water bath or fryer/cooker and the home thermostat. Can you name any other common devices or procedures that maintain homeostasis through negative feedback?

Observations:

Conclusion:

QUESTION: In homeostasis, there also are positive feedback mechanisms. Based on the definition of negative feedback, what is positive feedback? Can positive feedback mechanisms help to maintain homeostasis?

Observations:

Conclusion:

Journal Activity

One activity that will help you better understand and appreciate the human body is to keep a journal or log of your daily experiences throughout this course. The journal also may be a good location to record your ideas regarding the questions throughout this book. There are spaces beneath each question for your observations and conclusions, but you may have additional ideas when answering them. These ideas can be recorded in your journal.

Keeping a Journal

Objectives:
- To better appreciate the complexity of the human body.
- To become aware of your feelings and variety of experiences.
- To learn how to carefully and consistently record activities in a journal or log.

Materials:
- A journal or small notebook.
- A pen or pencil.

Procedure:

1. Obtain a small notebook or diary you feel will be convenient to use.

2. Select a time each day to record certain items that will be specified by your instructor. These may include your general feelings of health during the day, your oral temperature, your food intake, any medication you may take, and any side effects from the medication. If you become sick, also record any additional medication you take, how you feel during the illness, any changes in body temperature, length of illness, and so on. If you regularly exercise, record the type and duration of exercise, as well as how you feel during and after your workout.

3. The contents of your journal will be private and personal. Feel free to record this information. By studying the recorded information, you should better understand how your body functions.

Handling Body Fluids

Some exercises in this course may require you to obtain a small amount of blood by finger puncture or to use saliva for an analysis. The following are precautions and guidelines for handling body fluids, secretions, and blood. Please read these carefully, and follow them at all times. The greatest risk is associated with blood, because many diseases can be transmitted by exposure to it (i.e., blood-borne diseases). These precautions can, however, help to prevent the spread of disease from any source.

Universal Precautions for Collecting and Handling Body Secretions and Blood

Always wash your hands with the antibacterial soap provided before collecting from yourself or a partner. Gloves should be worn whenever blood, blood products, or body secretions are handled.

Occupational Safety and Health Administration (OSHA) standards require special equipment (e.g., goggles, acrylic shields, etc.) for handling human blood other than your own. To avoid these expensive requirements and reduce the chance for disease transmission, students should not handle someone else's body fluids. Gloves should be worn when handling equipment that may have come into contact with another student's blood. A person can collect his or her own blood by setting the capillary tube on the edge of the bench, thus freeing a hand to squeeze or "milk" the finger.

To collect blood, clean the area to be lanced with an alcohol swab and allow it to air dry. Do not blow on your finger to make it dry faster. Then, open the new lancet at the proper end by tearing the paper or foil covering where indicated or by twisting off the plastic shield, and make a quick puncture in the cleansed area. Lancets penetrate to a depth sufficient to allow adequate blood flow to fill several capillary tubes of blood. Several automatic lancing devices are available in the lab, for which specific directions will be given.

Gently squeezing or "milking" the finger may be necessary to obtain adequate flow. Cold hands may indicate decreased blood flow. Massaging the hands or swinging the arm in a circle will increase flow. Do not squeeze or milk too aggressively, however, because this will cause bruising and, later, soreness at the puncture site. Because of possible soreness, it is advisable to pick a finger on your nonwriting hand. Discard the lancet; *never reuse a lancet, even for yourself.* Gently squeeze the finger to collect the blood in a capillary tube, which will load automatically if held horizontally or slightly below the horizontal, or place the blood directly on a slide as needed.

Place all used materials with blood on them in the designated bag or biohazard disposal container. Do not handle another person's blood. Use gloves while cleaning up your work area, and wash hands both after handling the items used and at the end of the lab.

Certain reusable items will come into contact with blood, blood products, or diluted blood and body secretions. Put these in the designated containers, making sure that blood, blood products, or diluted blood do not come into contact with any equipment or surfaces unnecessarily. Clean these contacted surfaces and equipment with 1:10 fresh-bleach solution or a supplied cleaning/disinfecting agent. Clean your areas of benchtop at the end of the lab even if there is no suspicion of contamination. Keep all unnecessary items, such as books, purses, and coats, out of the way so they are not contaminated. Never eat or drink while body fluids are being collected, processed, or tested; it generally is good laboratory practice *never* to eat or drink in the same setting where body fluids are handled.

Section 2

Cells and Tissues

Note to the Student

When we begin study of a new subject we often start with the small, or micro, dimensions and work our way to the larger, or macro, dimensions. Therefore, we now start with the cells and tissues before tackling the systems of the body. To study this small world, we need to use the microscope. This section discusses the parts of a microscope and how they work. The other tools needed to study cells and tissues are slides. This section also explains how the slides you will use are prepared. Hopefully, this will allow you to approach this dimension of study with an appreciation of the tools you will be using.

Use of the Microscope

While you will not use a microscope extensively in this class, you should have some basic knowledge of its components and how they work. A microscope is one of the essential tools of introductory biology. Long before students are introduced to more complicated techniques and equipment, they are intrigued by descriptions and pictures of objects and structures that are too small to be seen by the naked eye. Microscopes allow us to enter that small world. From this journey, we get information that helps us to understand the smaller building blocks of structures that are visible to the naked eye and the details that help us to determine their function. Microscopic examination and study, however, is just one of many ways to decipher the anatomy and physiology of living systems.

Microscopes are composed of several essential components, and many other features can be added to improve functionality and ease of use. First, microscopes must have lenses to collect the light waves that pass through or reflect from the object being viewed (i.e., objective lenses). Second, they must have lenses to collect and deliver the light waves into our eyes (i.e., eyepiece or ocular lenses). There also must be a source of light (i.e., the mirror or light bulb) and a place to hold or set the specimen (i.e., the stage). The better the design and integration of these components, the better the microscope. Refer to Figure 2.1, learn the component parts, and get used to referring to those parts by their proper names. Now read the rules dealing with the handling and use of the microscope. Please follow these rules, because in an introductory class, the microscope can be the most expensive piece of equipment that you use.

Figure 2.1
A Microscope.
The microscope is an important tool that allows us to see the cellular components. Numerous types of scopes are available, but all have similar components.

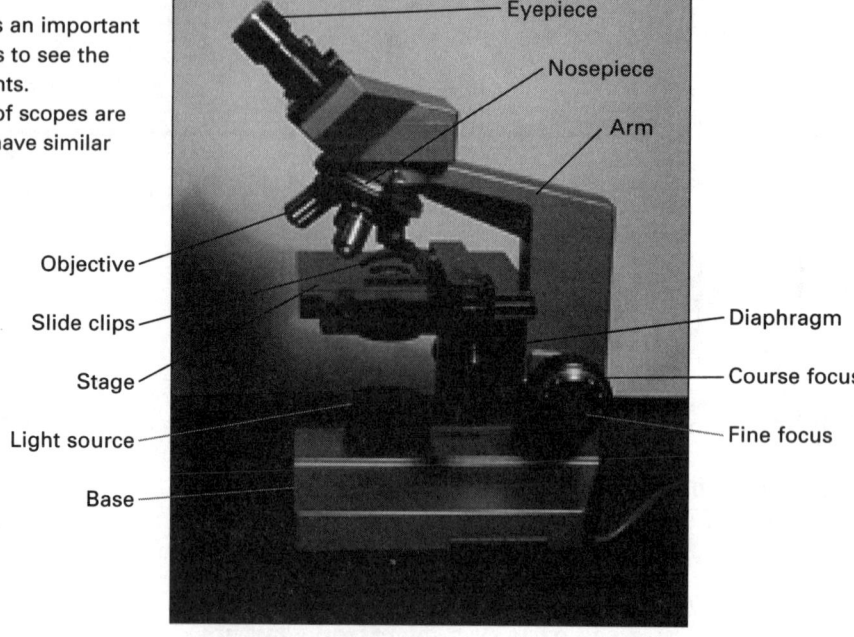

Use of the Microscope

Objectives:

- To learn the basic components of the light microscope.
- To understand aspects of the light microscope that can be adjusted to improve the image observed.
- To use a microscope for observing and studying objects too small to be seen with the naked eye.

Materials:

- Microscope.
- "e" slide.
- Lens paper.

Procedure:

1. Always handle the scope with two hands. Place one hand under the base and the other around the arm (Figure 2.1).

2. Set the scope squarely on the table and several inches away from the edge.

3. Before moving or using a scope, always make sure the nosepiece is set with the lowest-power lens in place.

4. Clean the "e" slide to be viewed, and place it on the stage. Secure the slide with the stage clips or within the spring-loaded arm of the mechanical stage. An "e" slide is a common tool for practicing operation of the microscope.

5. Begin focusing with the smallest lens on your scope. Adjust this lens to be as close to the slide as possible. Depending on the design of the scope, focus by raising the lens or lowering the stage until the slide comes into focus.

6. The slide should come into focus readily because lower-power lenses have the greatest depth of focus, which refers to how thick a "slice" of the specimen is in focus.

7. Adjust the light level. Usually, there are three ways to manipulate this. First, there often is a rheostat on the light control that allows the light level to be adjusted.

Realize that the color quality of the light changes based on the rheostat setting. Second, there typically is a substage diaphragm. Often controlled with a ring or a lever, it opens and closes a concentric, circular iris diaphragm to control the light level. Lower-power lenses need less light than higher-power lenses. Third, there may be a substage condenser that gathers and condenses the light-source output and concentrates it at the specimen. A common mistake is that students assume the light must be as bright as possible to see anything; too much light washes out the stain colors and reduces contrast.

Note: You may observe dirt on the eyepiece or lens. To tell if the dirt is on the eyepiece, rotate the eyepiece while looking through the scope. If the dirt moves, it is on the eyepiece. If the dirt does not move when the eyepiece is rotated, move the slide. If the dirt still does not move, it is on the lens. With a piece of lens paper, clean the dirt from whatever surface it is on. Do not use ordinary tissue paper on any glass surface of the microscope, and use special care when cleaning any lens surface. Your instructor may have particular guidelines for dealing with dirt on the microscope.

---

8. Most of the better microscopes are parfocal, which means that once the scope has been focused at one power, it can be switched to another power and still be essentially in focus. Therefore, when you are ready to switch to a higher power, simply rotate the nosepiece into place, then use only the fine-focus knob to adjust the focus.

9. If the scope is equipped with an oil-immersion lens and it is to be used, follow this procedure: After focusing to the high-dry lens (the 40X or 43X lens located before the oil-immersion lens) and as you prepare to switch to the oil-immersion lens, place a drop of immersion oil on the slide where you have been observing the section. Then, flip the oil-immersion lens into the drop of oil. As you move the slide on the stage, the oil will follow the lens. The purpose of oil on the immersion lens is to reduce the diffraction of light, thereby increasing the resolution.

10. When finished using the microscope, always leave the scope with the lowest-power lens in place. Remove the slide from the scope, and return it to the designated storage area. Finally, clean the lenses, especially the oil from the oil-immersion lens and stage.

Slide Preparation

The study of cells, or **cytology** (*cyto, cyte* (G), a cell; *logy* (G), study of), and the study of tissues, or **histology** (*histo* (G), tissue; *logy* (G), study of), depends on being able to observe structural details and chemical characteristics of cells and tissues. To accomplish this, both fields of study depend heavily on microscope slides.

Microscope slides are made by collecting tissue to be made into a slide. Either chemical treatment or heat is used to fix the tissue (i.e., to prevent its changing from its naturally appearing state). The tissue then is placed in melted paraffin and allowed to sit so that it becomes impregnated with the paraffin. This helps to maintain the structural integrity of the tissue during sectioning. The paraffin-impregnated tissue is placed in a mold, which often is referred to as a boat because of its shape. The mold is filled, and the tissue specimen is covered with paraffin. After cooling, the solidified boat is knocked out of the mold, and the excess paraffin shaved off. At this point, the tissue is supported by a pedestal of paraffin but is exposed on all sides (except where supported by the paraffin). The paraffin is melted to a wooden block, which provides a solid attachment point for the microtoming.

A **microtome** is a device that makes consistent, thin slices of the tissue. The microtome cuts either with a reusable metal blade that can be sharpened as needed or with broad "razor blades" that are disposable when dull. The tissue is mounted, and then every time the microtome wheel is turned, its head moves up and then out the set distance for the slice thickness desired. As the wheel is rotated, the head comes down across the blade, and a slice is produced. Because of the paraffin, the section adheres to the blade. The head goes up again, moves out the set distance, and drops down on the blade. Soon, there is a "ribbon" of slices or sections streaming off the blade. The ribbon of sections is removed from the blade and placed in a float or specially designed water bath, in which the sections separate and float. The individual sections then are placed on a slide, to which some "glue" (e.g., egg white) has been applied.

The slides with sections are heated and allowed to dry, after which the sections are ready to be stained. Staining is necessary, because there is little contrast between the cellular organelles. Staining allows parts to be differentiated. The slides are placed sequentially in bottle after bottle of stain, destain, and so on, depending on the specific staining procedure being used. They are then cleared, which is the process through which any water still present is removed from the section. Finally, a cover slip is attached with a transparent, nonyellowing mounting medium.

The slide is finished after it has set for several days on a slide warmer. The slide is examined and graded, and only at this point are previous errors detected. A common error that may require the slide to be discarded is an **artifact,** which refers to features on the slide that are not characteristic of the real, living tissue. Spaces, bubbles, and bent or distorted fibers may be artifacts depending on the particular tissue. Extreme artifacts require the slide to be discarded.

This process is not complicated, but it is time intensive. Treat slides with care. Only pick up one at a time and carry it in a cupped hand. Never keep a pile of slides on your desk, and never place a slide anywhere but on the stage of the scope or in its slide box. Do not set slides on your book. They are transparent and often are closed up in books as if they were bookmarks. Clean the slide before use with lens paper. Do not squeeze very hard, however, because the mounting medium never actually dries completely in the center of the slide. Also, avoid using solvents on a slide, because most mounting media are organic materials. If oil is present, use a small quantity of xylene to remove the oil while trying to stay away from the media.

Cytology: The Study of Cells

All living organisms are made of microscopic units called **cells.** The cell is considered to be the structural and functional unit of living things (Figure 2.2). Thus, learning the principal components of a typical cell and considering the processes carried out by these components are valuable. Cells are surrounded by barriers or cell membranes, and they contain a liquid or cytoplasm that in turn contains subcellular components or **organelles** that carry out the different jobs necessary for proper cellular function.

The Plasma Membrane

The outermost limits of a cell are determined by the **plasma membrane** (Figure 2.2). Also known as the cell membrane or the plasmalemma, the membrane is composed of proteins and lipids. Early attempts to construct a scientifically sound model of the lipid and protein arrangements in the plasma membrane were faulty because of inadequate knowledge about the type of lipids in the membrane and the jobs played by the proteins. Today, we know the plasma membrane is essentially a lipid bilayer containing **phospholipids** and **cholesterol. Proteins** are inserted into the membrane in a mosaic pattern. The name of the current model for the plasma membrane is the **fluid mosaic model.** While structurally important in the way they are inserted into the mem-

Figure 2.2
A Generalized Animal Cell. The cell, which is the structural and functional unit of all organisms, contains subcellular structures called organelles. Cells are specialized to carry out specific functions throughout the body and, therefore, have specialized shapes and other modifications. Any cell will have some or all of the illustrated organelles.

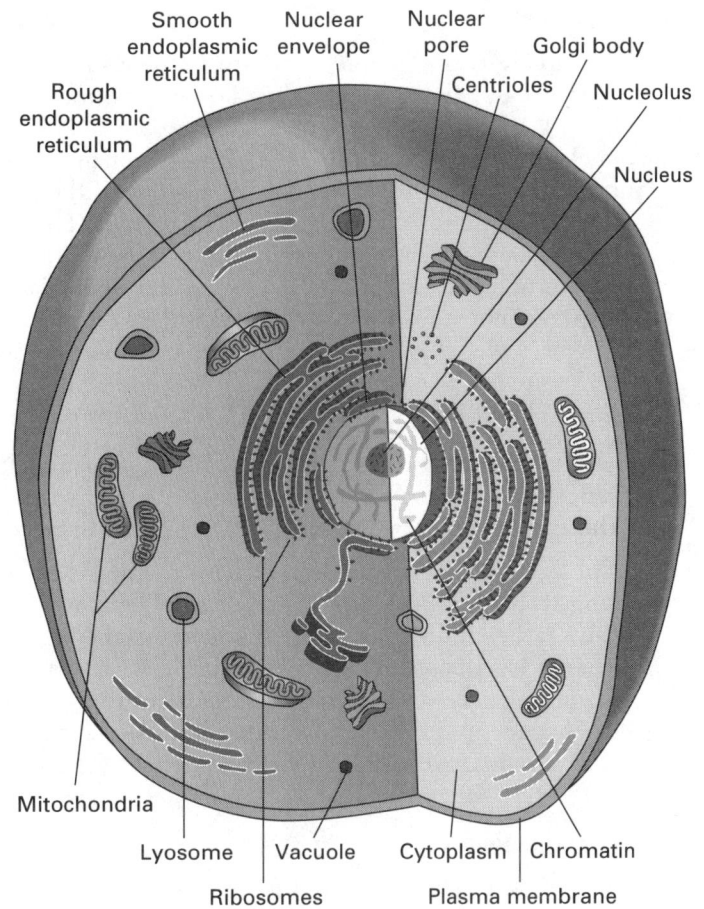

brane, the proteins function as receptors, markers, enzymes, pores, and gates in the cell membrane. Therefore, lipids provide the core or barrier function of the membrane, and proteins are the functional components of the membrane. This arrangement is referred to as a **unit membrane.** A good analogy is a wall of a house or a building. The blocks or wood and sheetrock are the lipid bilayer. They provide the barrier between the inside and outside or between rooms, but what can happen in the two separated spaces without some functioning components in the wall? Doors, windows, house numbers, and signs represent the proteins in this analogy. They allow the wall to function by controlling what and at what rate different materials get through.

Plasma membranes are selectively permeable, or **semipermeable.** This means that water moves readily through the membrane, but that the membrane can control the movement of most other components either inside or outside the cell. Lipid is not water soluble, so it is a good material to separate the watery environments inside and outside the cell. Only lipid-soluble materials can go through the cell membrane readily. Everything else must go through the pores and gates of the cell membrane; therefore, the membrane can control what gets into and out of a cell.

The Cytoplasm

The **cytoplasm,** or the transparent, elastic fluid with dissolved materials and suspended particles, fills the inside of the cell (Figure 2.2). When describing what the cytoplasm contains, we could use the analogy of a good old-fashioned soup. Such a soup might contain a little bit of just about anything in the kitchen. The vegetables floating in the soup would be the organelles, and all of the spices and seasonings would be the countless dissolved substances in the cytoplasm. It is not unreasonable to guess that the cytoplasm contains a bit of everything a cell likely needs to function.

The Nucleus

The **nucleus** is the largest and most obvious internal structure of cells on a slide of stained tissue (Figure 2.2). The nucleus takes most common stains well and tends to be dark. (Therefore, the dark, circular structure inside the cells you may observe is the nucleus.) It is surrounded by a double unit membrane–bound structure called the **nuclear membrane.** If we use our analogy of the house wall as the unit membrane, then the nucleus has two house walls around it. There are breaks in the nuclear membrane called nuclear pores, which are passageways between the contents of the nucleus, or **nucleoplasm** (or karyoplasm), and the cytoplasm. In the nucleoplasm, there are dispersed **chromosomes,** or **chromatin,** and one or more structures called **nucleoli** (singular, *nucleolus*), which are non-unit membrane–bound structures. The nucleolus is the site of ribosome synthesis. (Ribosomes are described later.) The chromosomes (i.e., colored bodies containing DNA and histone proteins) in the nucleus are the repository of genetic information for the cell's structure and function.

The Endoplasmic Reticulum

The **endoplasmic reticulum (ER)** is a unit membrane–bound set of channels running throughout the cell (Figure 2.2). There are two types: **smooth** or **agranular (SER)** when it does not have ribosomes attached to it, and **rough** or **granular (RER)** when it has ribosomes associated with it. RER is associated with the process of making proteins for release from the cell. The attached ribosomes "read" the **messenger RNA** made in the nucleus and that has entered the cytoplasm via the nuclear pores. SER is associated with lipid synthesis, either for the cell or for secretion.

The Ribosomes

Ribosomes are non-unit membrane–bound structures made in the nucleolus (Figure 2.2). They pass out of the nucleus via the nuclear pores and enter the cytoplasm. They are composed of proteins and ribosomal RNA and are the protein-manufacturing entities of the cell; this process is known as translation. There are two types of ribosomes: attached, and free. Attached ribosomes have already been mentioned in reference to the RER; free ribosomes make proteins for the cell's own metabolic needs.

The Golgi Apparatus

The **Golgi** or **Golgi apparatus** is a unit membrane–bound structure in the cytoplasm that is the packaging and shipping component of the cell (Figure 2.2). The proteins made by the RER enter the RER channels and are transported through these tubes to the Golgi apparatus. Any modifications to the proteins, such as removal of an extra segment or addition of a carbohydrate or lipid component, occur in the Golgi. The proteins then are packaged in vesicles for secretion from the cell.

The Mitochondria

Mitochondria are double unit membrane–bound structures in which the outer mitochondrial membrane is unfolded and the inner mitochondrial membrane is highly folded to increase its surface area (Figure 2.2). The space enclosed by the inner membrane is called the **matrix;** the folds of the inner membrane are called the **cristae.** Mitochondria are associated with the process of energy "manufacture" for the cell; energy-rich compounds are broken down and their chemical energy packaged in a form that the cell can use, such as adenosine triphosphate (ATP). Mitochondria are unusual in that they contain a small amount of DNA, which contains the information necessary to make more mitochondria. Thus, mitochondria are self-replicating.

Other Unit Membrane–Bound Structures

Lysosomes are unit membrane–bound sacks of hydrolytic enzymes that serve as digestion sites for the cell. Cells that are phagocytic contain numerous lysosomes. These cells phagocytose eat some cellular debris or invading organism to form a **phagosome**. Lysosomes fuse with the phagosome to carry out destruction of the ingested material. Much of the tissue damage following an injury is not a direct result of the injury itself but of the broken lysosomes destroying adjacent structures.

Peroxisomes are unit membrane–bound sacks of powerful oxidizing enzymes called oxidases. They function to detoxify alcohols, formaldehyde, and free radicals, which are highly reactive chemicals. These are converted to hydrogen peroxide, which subsequently is broken down to water by catalase.

Cilia are short, plentiful, unit membrane–bound extensions of the cell membrane. Because of their internal structure, they provide a propulsive force for either the cell or materials resting on the surface of the cell. Most cells do not have cilia. There usually are many cilia on cells that have them, however, and in humans, they line the respiratory passageways and help to move dirt-laden mucus as it is carried out of the respiratory system. Along with flagella, they show a remarkable consistency of internal structure, which is referred to as a 9 + 2 arrangement of microtubules.

Flagella are long, singular, unit membrane–bound extensions of the plasma membrane. They propel the cell. In humans, they are only associated with sperm cells, in which the flagellum is called a tail.

Other Non-Unit Membrane–Bound Structures

Centrioles contain an internal arrangement of microtubules of nine triplets. They are located outside the nucleus and serve an organizing function in the process of cell division.

The **cytoskeleton** is an internal mesh or scaffolding that provides structure for the cell and a place to attach the cytoplasmic inclusions. It is composed of small-diameter microfilaments, intermediate-diameter filaments, and large-diameter microtubules. The cytoskeleton helps to explain how cells maintain their shape and internal arrangement of organelles as well as how cells divide and move.

An Atlas of Tissue Types

One discovery that catches many students off guard is that for all the complex cells and tissues in the human body, there are only four tissue types. In other words, all different tissues can be grouped into one of four categories. These are: (1) connective tissue, (2) muscle tissue, (3) nervous tissue, and (4) epithelial tissue. We all have some notion of things that would be considered to be connective in nature as well as an understanding of what muscle tissue and nervous tissue refer to. Therefore, everything else that does not neatly fit into one of the other categories is epithelial tissue, thus making them very important to the overall structure of the body.

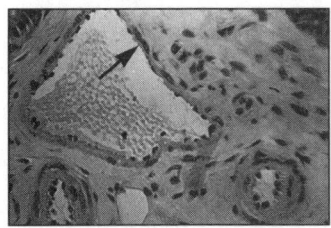

Figure 2.3
Simple Squamous Epithelium.

Epithelial Tissue

Epithelium usually is divided into two main categories: lining and covering epithelia, and glandular epithelia. The type that is most important to appreciate at the beginning of a course like this is the lining and covering epithelia. All epithelia, however, have certain common features. They do not have blood vessels running through the tissue. As a result, all epithelia are thin, because they depend on the underlying blood vessels to supply their needs. They also are attached to an underlying **basement membrane,** which is noncellular and contains connective tissue. It should already be

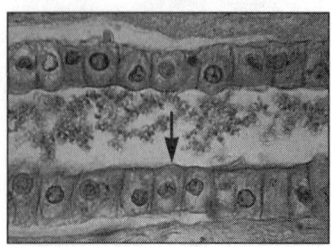

Figure 2.4
Simple Cuboidal Epithelium.

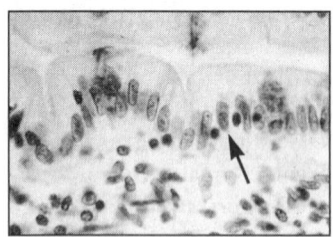

Figure 2.5
Simple Columnar Epithelium.

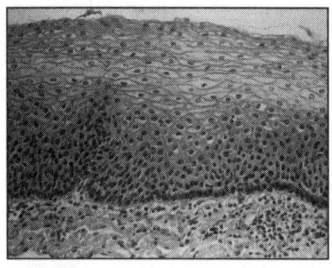

Figure 2.6
Stratified Squamous Epithelium.

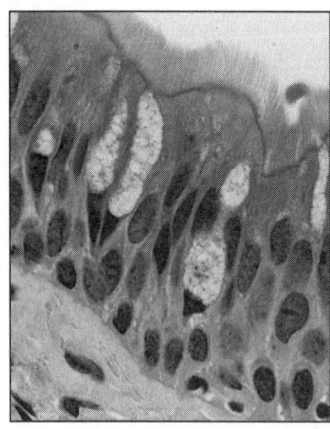

Figure 2.7
Pseudostratified Columnar Epithelium.

apparent that it is impossible to talk about one tissue type without mentioning the others. Rarely can you look at one tissue and see nothing else. Thus, epithelia are referred to as membranous, cellular tissues that cover and protect a surface, line a cavity, or produce a secretion.

Epithelia are described by the shape of the cells, whether the cells are in a single layer or multiple layers, and whether they have other components, such as cilia, goblet cells, or keratin.

Types of Lining and Covering Epithelia

1. **Simple Squamous Epithelium** (squama (L), a scale): The cells of this tissue are arranged in a single layer, and they are flat or scale-like in cross section (Figure 2.3). The nucleus is flat like the cell, but it often appears as a bump in the flat cell. They line body cavities and blood vessels, and they make up the air sacs in the lungs.

2. **Simple Cuboidal Epithelium:** The cells of this tissue are arranged in a single layer and are cuboidal in cross-section (Figure 2.4). The nucleus is located essentially in the center of the cell. Many glands and their ducts are lined with cuboidal epithelia.

3. **Simple Columnar Epithelium:** These cells are taller than they are wide, and they are shaped like columns in cross-section (Figure 2.5). They also are arranged in a single layer. Typically, the nucleus is toward the end of the cell that is attached to the basement membrane; this end of the cell is referred to as the basal surface of the cell. The digestive system and parts of the respiratory system are lined with this type of epithelium.

4. **Stratified Squamous Epithelium:** The cells of this tissue are arranged in many layers, and they cover surfaces exposed to the outside of the body (Figure 2.6). This tissue is found in areas experiencing a lot of wear and tear. Thus, the many-layered arrangement provides protection for the underlying structures. This tissue is what makes up the epidermis of the skin, and in this case, these cells contain a protein called **keratin**. Hence, the epidermis is a keratinized, stratified squamous epithelium. The mouth, pharynx, esophagus, and vagina are lined with the nonkeratinized version.

5. **Pseudostratified Columnar Epithelium:** In this type, the cells appear to be arranged in layers, because the nuclei of the columnar cells are not in a row toward the basal surface of the cell (Figure 2.7). Each cell is in contact with the basement membrane, however, which is not the case in a true stratified epithelium. Hence, this tissue is falsely stratified or pseudostratified. It often has cilia on the free or apical surface and thus is referred to as pseudostratified, ciliated columnar epithelium. It lines large portions of the respiratory tract.

6. **Transitional Epithelium:** This is a stratified epithelium that is specialized for stretch and expansion (Figure 2.8). It is found lining the urinary bladder and associated tubes. Its free surfaces are not made of squamous-shaped cells but of irregular to cuboidal cells that do not separate and tear when stretched.

There are other, rarer types of epithelia, but they are best treated as an academic rather than a practical consideration.

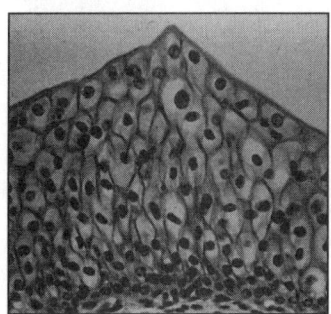

Figure 2.8
Transitional Epithelium.

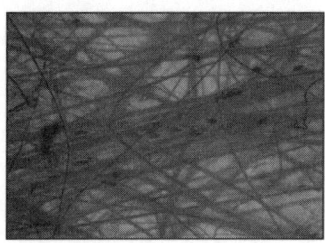

Figure 2.9
Loose or Areolar Connective Tissue.

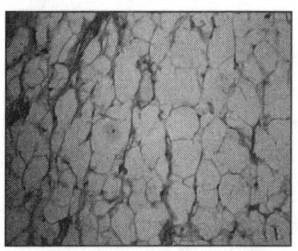

Figure 2.10
Adipose Connective Tissue.

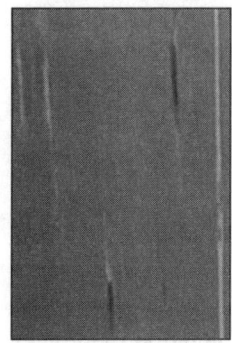

Figure 2.11
Dense Connective Tissue.

Connective Tissue

Connective tissue includes bone and cartilage, which come to mind immediately when we think of a connective function. It also includes tissues such as loose, dense, and reticular connective tissue; adipose tissue; and blood. That blood is indeed considered connective tissue is another small surprise for some. What, then, are the unifying features that make connective tissue a coherent assemblage?

The single most important feature of connective tissue structure is referred to as the extracellular matrix. These tissues generally have a large amount of extracellular material that is noncellular. In other words, there is a lot of space between the cells of connective tissue. Indeed, the matrix contains fibers that are typically straight fibers of **collagen,** which are referred to as collagenous or white fibers; stretchy fibers of **elastin,** which are called elastic or yellow fibers; and branching collagenous fibers, which are called reticular fibers. The **fibroblasts** that secrete the matrix fibers and the **macrophages** that clean up and protect are some of the cells found in connective tissue. There is some variation in the vascularity of connective tissue as well, from no blood supply at all, as in cartilage, to the blood itself. Most types of connective tissue are discussed and described here, but others, such as bone and blood, are discussed later.

Types of Connective Tissue

1. **Loose or Areolar Connective Tissue:** There is much open space in loose or areolar connective tissue (Figure 2.9). The space is not empty, however, but open, implying that cells are not in direct contact with other cells. The matrix contains all three fiber types and serves as basement membranes and nets around blood vessels and organs.

2. **Adipose Tissue:** When seen on a typical microscope slide, adipose or fat tissue has large, open areas (Figure 2.10). This results because in the living tissue, adipose cells or adipocytes normally are full of lipids in the form of **triglycerides.** During slide preparation, the solvents wash these lipids out of the cells, leaving large, artifactual holes.

3. **Dense Connective Tissue:** Dense connectve tissue can be divided into two subcategories: regular, and irregular (Figure 2.11). The dense, regular connective tissue is more compact than either of the types already mentioned, and it is characterized by all collagenous fibers running in the same direction, as occurs in tendons and ligaments. The irregular form contains collagenous and reticular fibers that form a tight mesh around bones and cartilage and in the dermis of the skin.

These three types of connective tissue fall into what is referred to as **connective tissue proper.** This is in contrast to the more specialized, well-known tissues of cartilage and bone. Cartilage is considered here, but bone as a tissue is considered in Section 4, Bone and the Skeletal System.
Cartilage

Cartilage is characterized as having a gel-like material with the protein fibers in the matrix. Cells that secrete cartilage are called chondrocytes. Because they secrete the matrix, they "paint themselves into a corner" until they are surrounded by that matrix. The spaces that remain where the cells are located are called **lacunae.** Remember that cartilage has no blood supply, so do not expect to see any blood vessels in a slide of cartilage.

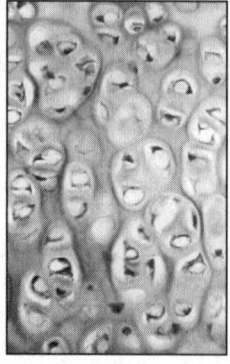
Figure 2.12
Hyaline Cartilage.

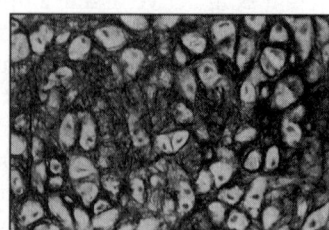
Figure 2.13
Elastic Cartilage.

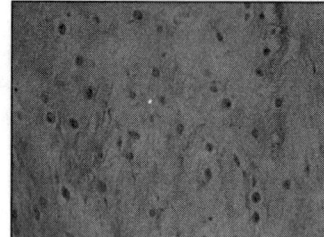
Figure 2.14
Fibrocartilage.

QUESTION: If you have ever sprained a part of your body or torn a ligament, you know these types of injuries are slow to heal. What do you know about connective tissue in general, and cartilage specifically, that might affect its rate of healing?

Observations:

Conclusion:

Types of Cartilage

1. **Hyaline Cartilage:** Hyaline cartilage has a uniform and smooth matrix, hence the name hyaline ("glassy") (Figure 2.12). It is found as reinforcing rings in the respiratory tubes, cushioning at the ends of bones, and at the ends of ribs.
2. **Elastic Cartilage:** In elastic cartilage, the matrix has distinct elastic properties because of its elastic fibers (Figure 2.13). Elastic cartilage is found in the external ears, the nose, and the epiglottis.
3. **Fibrocartilage:** Fibrocartilage is characterized by thick, collagenous fibers that are obvious in the matrix (Figure 2.14). It is found in the intervertebral discs, the cushions (i.e., menisci) in the knees, and in the pubic symphysis.

Diffusion and Osmosis

Later, we will discuss the circulatory system, which delivers materials over long distances in the body. It picks up the products of digested foods at the intestines and oxygen at the lungs, and it carries these to cells all over the body. The blood then picks up metabolic wastes and carbon dioxide to be carried away from the cells. The force for moving materials over these long distances is the muscular pumping action of the heart, but the mechanism for pick up and delivery is **diffusion.** Diffusion is the process whereby materials, solids, liquids, or gases move from an area of higher concentration

Figure 2.15
Diffusion.
Diffusion is the process whereby materials move from areas of higher concentration to areas of lower concentration.

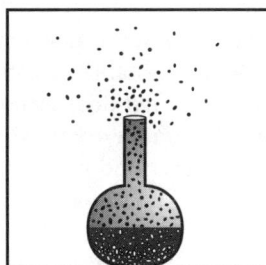
(1) High concentration of fluid molecules in bottle.

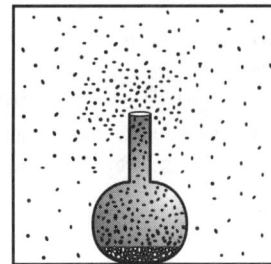
(2) Fluid evaporates and molecules move from where they are more plentiful to where they are less plentiful.

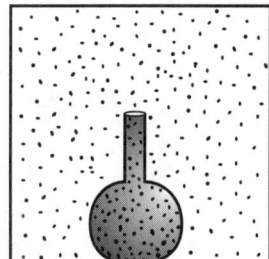
(3) Molecules distributed everywhere evenly.

to an area of lower concentration (Figure 2.15). In biological systems, diffusion often occurs from one compartment to another, such as the intracellular and extracellular spaces. These spaces are separated by semipermeable membranes. Diffusion is a **passive** process, because it does not require the addition of energy (in the form of ATP) for it to occur. Diffusion occurs based on the energy already in the system.

QUESTION: Every thing on Earth has energy associated with it, and we are aware of the energy levels in our environment. What do we call the average energy level in an environment? (Hint: We measure this parameter with a thermometer.)

Observations:

Conclusion:

Water is a very biologically important molecule, because life is based on the reactions that occur in an aqueous or watery environment. Being such a ubiquitous liquid, water also moves from areas of high concentration to areas of low concentration. Because water is so important to all organisms, the term **osmosis** is used to refer specifically to the movement or diffusion of water.

Diffusion and Osmosis

QUESTION: Water has properties associated with its chemical structure that help to explain much of its behavior. Water is very temperature-stable, resists freezing, and requires a lot of heat to turn into steam or water vapor. What does water do during freezing that is different from other liquids and causes damage to cells? (Hint: What happens to rigid containers when they are filled with water and frozen?)

Observations:

Conclusion:

In addition to freezing of water, excessive movement of water, based on the concentration of solutes in the aqueous solutions, also is damaging to cells. Water is very important, but cells cannot actively move water from areas of low to high concentrations. The process of moving materials against the concentration gradient is called **active transport,** and this process uses energy (in the form of ATP) to power the pumps that move materials "uphill." This does not occur for water. Water movement is vital to cells, however, so for cells to control water movement, cells must be water impermeable as well as able to set up water gradients by adjusting solute gradients. This means that if a cell needs to move water from one location to another, it can move solutes into the area in which water needs to be moved. In essence, this creates a lower water concentration, so the water will move from the higher concentration to the lower concentration (i.e., essentially down the gradient).

Observing Diffusion and Osmosis

The rate of diffusion is affected by the temperature of the system, the concentration gradient (i.e., how different the high and low concentrations are), the size of the molecules that are moving, the solubility of the molecules in the system, and, very often, whether the molecules are charged, uncharged, or lipid-soluble. Water movement can be measured with a device called an osmometer.

QUESTION: Speculate as to the effects of each item mentioned in the previous paragraph. How will they affect the rate of diffusion? (For example, what will a higher or lower temperature do?)

Observations:

Conclusion:

Diffusion

Objectives:

- To understand the process of diffusion.
- To appreciate the effects of certain factors on the rate of diffusion.

Materials:

- Petri dishes with agar.
- Cork borer or pipette to cut holes in the agar.
- Methylene blue solution, 0.1 M aqueous.
- Congo red solution, 0.1 M aqueous.
- Marking pen.
- Refrigerator.

Procedure:

1. Obtain three agar-filled petri dishes, or carry out the following procedure to prepare the dishes: Make a 1.0 to 1.5% agar solution. (Agar must be heated until near boiling to dissolve; be careful to avoid it boiling over.) Pour app. 25 ml in each dish, cover, and allow to cool. Sterile technique is not necessary, because the exercise will not last long enough to allow much contaminating growth. Also, because this is nonnutrient agar, there is not much for bacteria to feed on.

2. With a cork borer, a small plastic tube, or the end of a pipette, poke two holes or wells in the agar, then remove the plugs (Figure 2.16). The holes should be consistent in shape and size and no more than 1 cm in diameter. Place the holes equidistant from the walls of the dish, allowing as much distance as possible between the holes. Try not to crack the remaining agar or separate it from the petri dish surface.

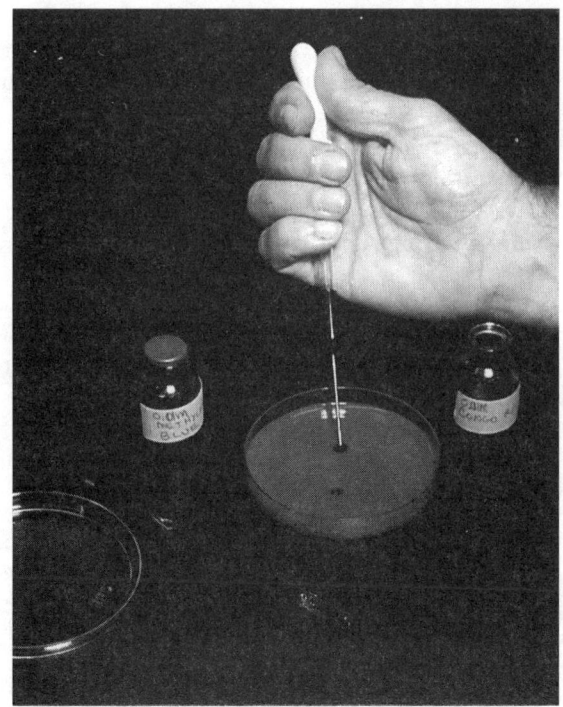

Figure 2.16
Preparation of Petri Dishes for Diffusion Exercise.
Small holes are cut into the agar to create wells, which are filled with stain solutions. Dishes are labelled and placed where they will not be disturbed for at least 24 hours.

3. Fill one hole with Congo red solution (0.1 M aqueous; MW, 698 g) and the other with methylene blue solution (0.1 M aqueous; MW, 319 g) in two of the three plates. Be careful not to overfill the holes or spill the contents of the holes when handling the dishes.

4. Label one plate as "control," and leave it on the classroom counter. Label the other as "refrigerator," and place it into a refrigerator.

5. Take a small subsample of the concentrated stains, and dilute the subsample in half with water. Now, fill the holes in the last dish as described earlier. Label this as "diluted," and leave it out on the counter as well.

6. The wells will empty quickly, but diffusion will continue for several days. The plates should be read at 24 hours by measuring the diameter or radius of the circles for each stain in the three dishes.

7. Realize that the effects of temperature, molecular size, and concentration gradient are being examined in this exercise. Consider each of these as you evaluate the results.

Because water is clear and colorless, its movement is hard to observe. In the following procedure, water movement is determined through the movement of fluid up a glass tube, which can be observed and measured over time.

Osmosis

Objectives:

- To differentiate the processes of diffusion and osmosis.
- To appreciate the importance of water and water movement in living systems.
- To appreciate the pressure generated by water movement.

Materials:

- Dialysis tubing.
- Rubber stopper, #5 with one hole.
- Heavy string.
- Glass tubing, disposable 5-ml pipette with end cut off or rigid plastic tubing.
- Large beakers.
- Glass marking pens.
- Six-inch rulers.
- Sucrose solutions, 10% and 30% aqueous.
- Clamp and ring stand.

Procedure:

1. Prepare two osmometers (Figure 2.17). Obtain similar-sized pieces of dialysis tubing that have been soaking in water. (Dialysis tubing is a cellulose-based material with microscopic holes that allow small molecules, but not large ones, to move through. It is packaged dry, so it must be soaked before handling so as not to crack it.) Open the tubing under running water. (The tubing comes as a tape and must be opened to become a tube.)

2. Twist one end of the tubing tightly, fold it over, and tie it securely with string. Insert the other end in an appropriate-sized, one-hole rubber stopper, and securely tie the tubing to the stopper. You now have a bag with a rubber stopper in the end.

3. Fill one bag through the stopper, using a large syringe with a large-gauge, blunted syringe needle, with one of the sucrose solutions. (Two sucrose concentrations probably will be available: 10%, and 30%) Avoid trapping bubbles in the bag. Fill the stopper with the solution, and insert a glass tube into its hole.

Figure 2.17
Osmometer Preparation.
An osmometer can be constructed from dialysis tubing, a rubber stopper, string, and a glass tube. The bag is filled with a concentrated sugar solution.

Caution: Exercise caution when inserting the glass tube or pipette into the stopper. Insert the tube just far enough to secure it in place, and lubricate the tubing with a small amount of liquid soap if it is difficult to insert.

4. Attach the osmometer to a ring stand with a clamp, then immerse it in a beaker of water so that the bag surface below the tie on the stopper is completely covered with water (Figure 2.18).

5. Mark the fluid level in the tube, and record the time. If there are no leaks in the system, the fluid should immediately begin to rise. A leak will look like a shimmering waterfall, streaming from the bag surface into the water.

6. Repeat the process to set up the second osmometer. Keep the osmometers set up as long as possible, and remember to record the time and mark the tube at set intervals. It is best to mark the tube every 10 minutes in case a leak develops and the fluid stops rising.

Figure 2.18
Osmometer in Operation.
The prepared osmometer is placed into a beaker of water and held in place by a clamp. The height of the column of fluid is marked, and the time is noted.

Observing Diffusion and Osmosis

QUESTION: In which tube did the fluid rise fastest? If there was a higher concentration of sucrose in one bag, which bag has the greatest concentration of water? (Remember there are holes in the bag and osmosis is the movement of water from areas of higher to lower concentration.)

Observations:

Conclusion:

Cellular Division

One thing that may be fairly obvious from the previous discussion is that many different types of tissues are found in the body. Remember that all human life starts from the fusion of a sperm and an egg, which together form a zygote. The zygote goes through many cell divisions to become an embryo, and then its cells continue dividing to produce a fetus. At each level of development, new tissues and, eventually, new organs and systems form, until all of the components are present. Eventually, after much growth, birth occurs. Humans are born with an estimated two trillion cells in our bodies. As we grow, we add cells to our bodies by the division of existing cells to make more. We also continually loss cells in the form of dead skin cells, red blood cells, digestive tract lining cells, and a host of other cell types. Cells for growth as well as cells for repair and replacement require a mechanism that allows cells to be produced.

Growth represents two different processes. The type of growth just alluded to—that of making more cells by division—is called **hyperplastic growth,** or **hyperplasia** (Figure 2.19). Between divisions, a cell typically grows to a predetermined size, until it is ready to divide into two new cells; this type of growth is **hypertrophic growth,** or **hypertrophy**. Therefore, the processes of growth and development really are a combination of hyperplasia and hypertrophy. Certain types of tissue, such as muscle and nerve tissue, lose the ability to grow via hyperplasia and are only capable of enlarging in response to increased demands. Most other types maintain the ability to grow by either mechanism.

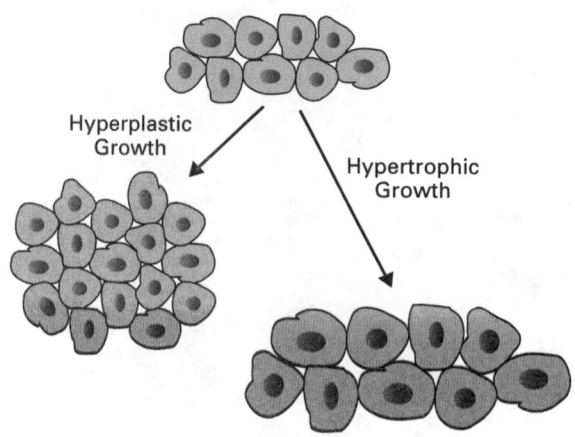

Figure 2.19
Hyperplastic versus Hypertrophic Growth.
Tissue and the cells they contain can grow in different ways. For overall growth, the tissue component cells must get larger, as in hypertrophic growth. To replace dead or dying cells, the number of cells must increase through hyperplastic growth.

The process of cell division often is called **mitosis;** however, this is an improper use of the terminology. Mitosis actually refers only to nuclear division or partitioning of the chromosomes into the two new cells' nuclear areas. Another name for mitosis is **karyokinesis** (*karyo* (G), the nucleus; *kinesi* (G), movement, moving). Therefore, subsequent division of the cytoplasm must occur if the process of cell division is to be complete. This is referred to as **cytokinesis.** Thus, the process of cell division is actually the process of karyokinesis or mitosis followed by the process of cytokinesis.

The Cell Cycle

Cells are not always dividing or preparing to divide. There often is considerable time when cells are carrying out the processes required to perform the functions of the cell. The term **cell cycle** refers to the time spent by the cell in its various activities (Figure 2.20). It is divided into phases. The majority of a cell's time is spent in **interphase,** which is the time between cell divisions. In turn, the interphase is divided into subphases. The G_1 **phase,** or **growth phase 1,** is when the cell is making the components necessary to grow and to function. The **S phase,** or **synthesis phase,** is when the cell's chromosomes are duplicated. The term synthesis refers to the making of new DNA for the new chromosomes. The cell then enters the G_2 **phase,** or **Growth phase 2,** during which the cell continues its growth, functioning, and development. At the very end of this phase, the cell prepares for the next cell division. The mitotic part of the cycle is further subdivided into phases based on obvious structural changes that occur in the cell during the process; the mitotic phases and their characteristics are described in Table 2.1. When the telophase ends, and cytokinesis occurs, the process of cell division is complete.

Figure 2.20
Cell Cycle.
Illustration of the cell cycle showing the phases a typical cell goes through after it divides and then grows or develops and prepares for the next division.

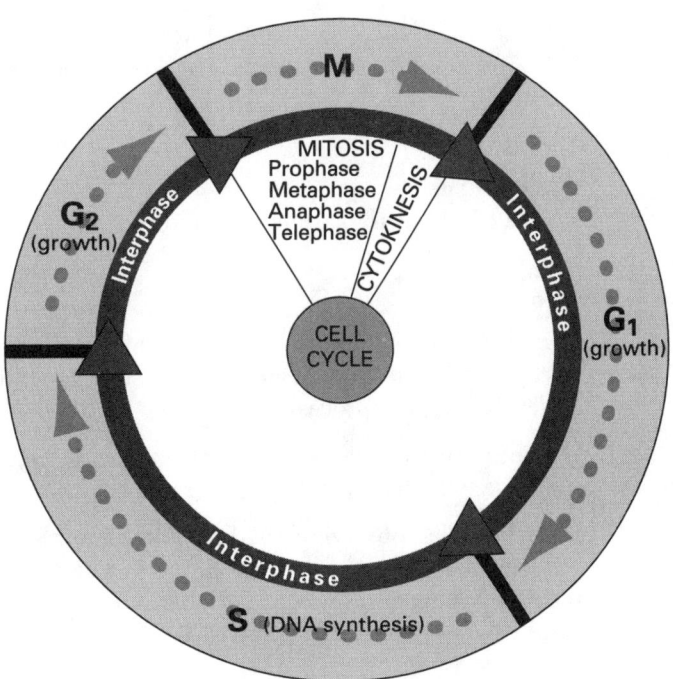

Table 2.1 The Stages of Mitosis

Prophase

Chromatin condenses into chromosomes
Nucleolus disappears
Centrioles separate
Mitotic spindle forms
Nuclear membrane disassembles

Metaphase

Chromosomes align on the metaphase plate

Anaphase

Chromosomes move to the opposite poles

Telophase

Chromosome movement stops
Nucleolus reappears
Mitotic spindle disassembles
Nuclear membrane reforms

Cell Division

Objectives:

- To understand the process of cellular division as composed of mitosis and cytokinesis.
- To use a microscope to study slides.
- To observe differences in the appearance of chromosomes to identify stages of mitosis.

Materials:

- Microscope.
- Slide of whitefish blastula or onion root tip.
- Lens paper.

Procedure:

1. Obtain a slide of whitefish blastula or onion root tip (Figure 2.21). Neither of these tissues represents human mitosis, but they do provide good study material for understanding the process.

2. Do not expect to see the stages of interphase, mitosis (i.e., prophase, metaphase, anaphase, telophase), and cytokinesis in any pattern or sequence on the slides.

3. Scan the slides, observing good mitotic "pictures." The term pictures refers to the different stages of the mitotic process, which are in stop motion on the slide. Identify what you observe, and continue scanning until representatives of each stage have been located.

Figure 2.21
Mitosis and Cytoplasmic Division.
This set of photomicrographs displays representative stages of mitosis and cytoplasmic division, such as prophase, metaphase, anaphase, and telophase.

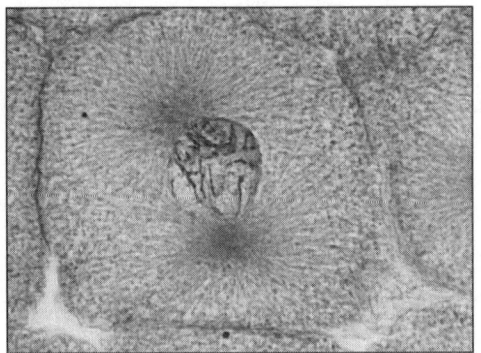
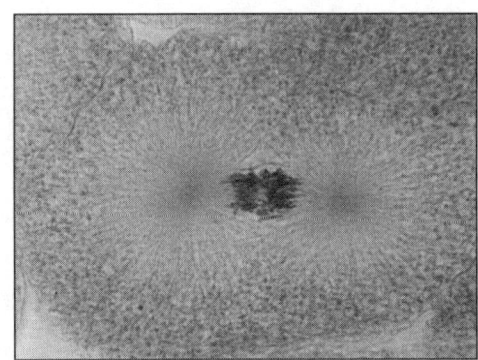
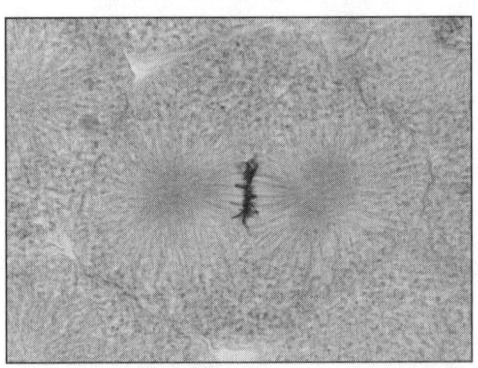
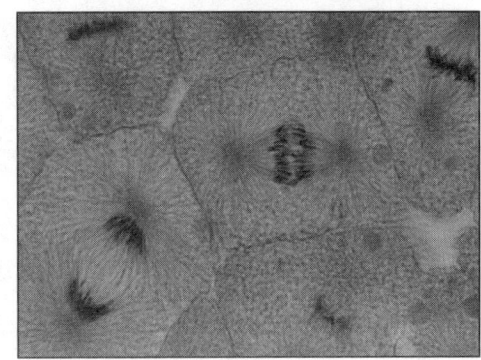
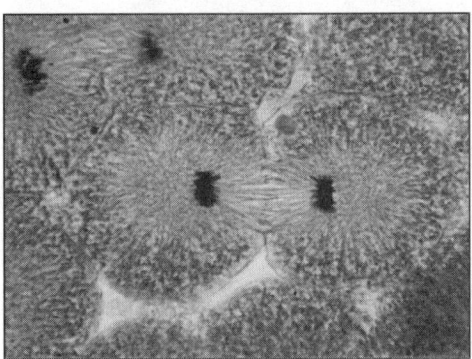

QUESTION: Cells divide for two reasons: to replace cells that are lost because of wear and tear or injury, and to make new cells for growth and development. What are some cells that are lost because of wear and tear? Think about body parts that undergo a lot of physical stress and strain and about cells that only live for a certain length of time. (Hint: What do you think happens to the lining of the digestive tract as you swallow a handful of popcorn?)

Observations:

Conclusion:

The Cell Cycle 31

Section 3

The Integument

Note to the Student

This section is a change from the more anatomical and histological treatment in the previous one. It includes several exercises that will help you to understand the truly dynamic nature of the skin.

Structure and Histology of the Skin

The **integument,** or the skin, is the largest organ of the body. It certainly covers the greatest area, and it is as extensive as any other system. The skin protects the body from external chemical, viral, and bacterial attacks. (There usually is considerable concern when treating injuries as to whether the skin is broken, because broken skin provides easy access into the body.) Protection from sunlight also is a function of the skin. Exposure to sunlight causes pigment cells to produce more melanin, which absorbs harmful radiation from the sun; this is what we call the tanning process. In addition, your skin is the site for many sensory organs and cells that detect stimuli in your environment. Hot and cold receptors, pressure receptors, touch receptors, and pain receptors are but a sample of the numerous specialized structures that allow us to detect changes in the world around us.

The only feature of the skin to study—as far as anatomy is concerned—is its microscopic structure. Skin is composed of two tissue layers (Figure 3.1). The outermost layer is the **epidermis** (*epi* (G), upon; *derma* (G), skin), and the underlying layer is the **dermis.**

Remember that the epidermis is an epithelial tissue layer; therefore, it does not contain any blood vessels and it depends on the underlying dermal layer for its nutritional supplies and waste elimination. Shallow cuts and scrapes that bleed may not seem to be very deep, but they must penetrate or damage the dermis to produce blood. The epidermis is made of sublayers that are based on the function or appearance of the cells.

The deepest layer of the epidermis (i.e., the layer next to the dermis) is the base layer, or the **stratum basale** (*stratum* (L), layer; *basa* (L), a base). It contains the single layer of fully functional, living epidermal cells. Being closest to the blood supply, these basal cells can rapidly divide through mitosis and provide new cells to maintain the integrity of the skin or to produce the cells necessary for repair following an injury. The **melanocytes,** or **pigment cells,** also are located in this sublayer. They produce melanin and, because of their star-shaped configuration, can distribute pigment to the basal cells.

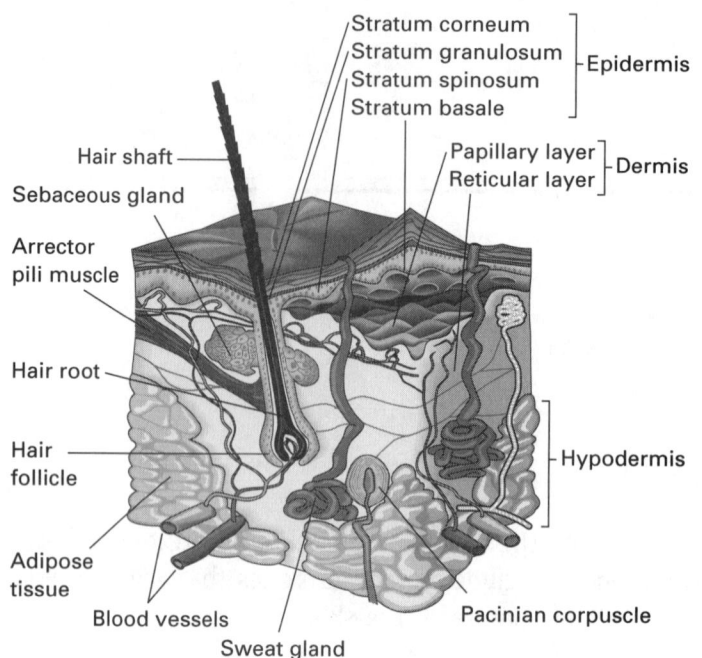

Figure 3.1
Diagrammatic Illustration of a Section of Human Skin.
Note the two principal layers: the dermis, and the epidermis. The epidermis is the protective, covering epithelium. The dermis is the nutritive, supporting layer that is rich in blood vessels and contains most of the epidermal structures that we associate with skin.

The next sublayer is the **stratum spinosum** (*spina* (L), a spine), so named because of the odd, spiney shapes of the cells in this sublayer. These cells, along with all the others in the skin, are being pushed to the surface of the epidermis through mitotic growth of the stratum basale. They are not as "alive" as the basal cells, and they show signs of decreased metabolic functioning.

The next sublayer is the **stratum granulosum** (*granul* (L), a little grain; *ose* (L), full of). These cells have a granular appearance, in part because of their increasing dehydration. **Keratin,** which is the tough protein associated with the skin, accumulates in these cells, which are less alive than the spinosal cells.

Next is the **stratum lucidum** (*lucid* (L), light). This sublayer only can be seen on the palms and soles of the body. In these two areas, the skin is thicker, and this layer is part of the protective function of the skin in these parts of the body. The cells are densely packed, and they have lost the individual cell boundaries between themselves, appearing as a light-staining, uniform layer.

The last sublayer of the skin is the **stratum corneum** (*corne* (L), horn, horny). These are the "dead" cells of the skin, which provide protection against the constant abrasion of the skin through our daily activities. They wear off and are continuously replaced by new cells from the basal layer. (Along with soap scum and body oils, these cells make up the "bathtub ring.")

Journal Note Fingerprints result from the dermal structural arrangement showing through on the surface of the epidermis. Your finger- and toeprints can by observed directly, but recording images of these prints makes their study easier. As we all know, fingerprints are recorded by pressing the surface of the fingertip on a stamp pad, then rolling the fingertip on paper. Record your prints, and observe the variety of patterns. Describe the patterns observed.

Journal Note Besides fingerprints, another pattern showing through from the underlying dermis are dermal ridges. There also are tension lines associated with the joints. To make these patterns more obvious for study, allow ink from a pen to run out on the skin surface. The ink will run in the grooves and between the ridges, thus allowing the lines and patterns to be observed.

QUESTION: After long periods in the bathtub or swimming pool, the fingers and toes appear to be wrinkled. Actually, these "wrinkles" are not what they seem. Can you explain why the skin appears this way?

Observations:

Conclusion:

The second layer of the skin is the dermis. This is a layer rich in blood vessels, subcutaneous fat, nerves, sense organs, lymph vessels, and muscles. Several structures of the epidermis also are found in the dermal layer. The **hair follicles, fingernails** and **toenails,** and the **sebaceous** and **sweat glands** actually are epidermal in origin, but they apparently need the anchoring and richer blood supply the dermal layer provides. All three of these structures extend through the epidermal layer, but they are best observed in the dermal layer. As you observe these structures on slides or models, remember the comments made earlier about sectioning techniques. You might have a fairly realistic mental image of the shape for each of these structures, but you are looking at slices through these structures. Depending on the angle of the hair follicles and sweat glands, they may take on very diverse shapes. Note that each hair follicle has a sebaceous gland associated with it. (If you do not see one, it just did not get cut in that particular slice.) The gland produces oils, which are called **sebum,** to lubricate and moisturize the hair and surrounding skin. Some people produce lots of this oil and have oily skin; others produce little and have very dry skin. More difficult to observe is the small muscle attached to every hair on the body. This muscle, or *arrector pili* (*arrect* (L), upright; *pilus* (L), hair), raises the hairs of the body when we are nervous, scared (this is what makes wild animals look bigger or more fierce), or cold (to hold more still air next to the body and conserve body heat). Sebaceous glands appear as clusters of cells draining toward the hair. Observe these well, and remember their appearance when you look at other glandular tissue later. Each sweat gland is composed of a long tube of cells, with the part in the dermis "balled up." A slice through this arrangement looks like a ball of string cut with scissors. Refer to your text for more information about the function of the skin and other details regarding the dermis and epidermis.

Journal Note   Hairs on different areas of the body have different textures and behave differently. Remove hairs from different parts of your body, and study them. If possible, observe the hairs under a microscope. Can you get a feeling for their cross-sectional shape? Do curly hairs look the same as straight hairs?

The Structure of Skin

Objectives:

- To observe the structure of skin.
- To identify some of the sensory structure in skin, if possible.
- To observe the epidermal structures in the dermis.

Materials:
- Microscope.
- Section of skin.

Procedure:

1. Place the slide of the skin on the microscope stage.

2. Observe that the skin consists of two layers: the epidermis, and the dermis. Note that there are no blood vessels in the epidermis.

QUESTION: You probably have had splinters stuck in your skin. Some splinters produce bleeding, but most do not. What does the observation of most splinters not producing bleeding tell you about how deep the splinters have penetrated?

Observations:

Conclusion:

---

3. Identify the layers of the epidermis: stratum basale, stratum spinosum, stratum granulosum, stratum lucidum (if the skin section is from the sole or palm), and stratum corneum. Note the many sublayers that make up the corneal layer, and how the cells on the surface are very flat and empty. Note also that the basal layer has dark pigment associated with it; this is melanin, the brown/black pigment of most animals.

4. Identify the papillary layer and the reticular layer of the dermis. If hair follicles are present, note the layered structure of the follicle. There should be a sebaceous gland associated with the hair as well. Can you see the arrector pili muscle attached to the hair follicle?

QUESTION: Each hair on the body has a small muscle attached to its base. Therefore, the hairs can be made to stand up. What are the two reasons why hair would stand up? (Hint: This is what causes "goosebumps" or "gooseflesh." What is this condition associated with?)

Observations:

Conclusion:

5. Can you see any sense organs in the skin section? These often are very small and difficult to see. If you observe any likely structures, refer to Figure 3.1 for identification. Sweat glands are large and should be seen easily.

Two-Point Discrimination

The purpose of this exercise is to demonstrate differences in the density of touch receptors at different parts of the body. As you might guess, the hands and feet are richly supplied with touch receptors. With these receptors and the muscle configurations at these parts of the body we can perform remarkable feats with our hands (and feet, as shown by some people without full use of their hands). The face is another area containing many receptors per square-centimeter. Most of the rest of the body will have fewer receptors and, therefore, greater spacing between receptors.

Two-Point Discrimination

Objectives:

- To understand that sensory fields of receptors are truly macroscopic in dimension.
- To realize there are different distributions of receptors throughout the body, depending on the sensory nature of each area.
- To learn how to do blind testing with reliable results.

Materials:

- Dividers or compasses.
- Small rulers.
- Graph paper.

Procedure:

1. Your task is to determine how far apart two points of contact must be before the subject can consistently and accurately detect two points of contact. To show this, obtain a divider or compass from the instructor. Select a spot on one fingertip, and touch the skin with the divider points (Figure 3.2). This test requires two persons: a subject, and an experimenter.

Caution: Compass points may be sharp. Before using any compass, clean the tips with an alcohol pad. Be careful as you touch parts of the body with these tips; laying the tips on the surface of the part touched will work.

Figure 3.2
Two-Point Discrimination Procedure.
Note the technique of touching the skin with the tips of the divider or compass. Vary the pattern of touching between one point and two points to "keep the subject honest."

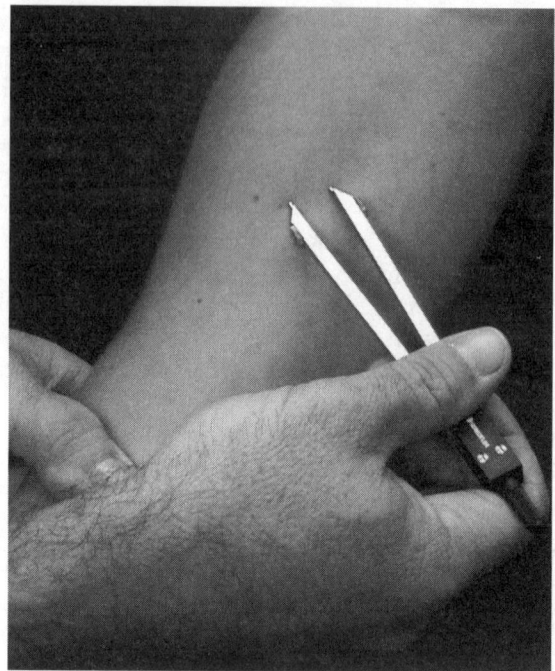

2. The subject should look away or keep his or her eyes closed so as not to receive visual cues about the procedure. Start with the divider points very close together, and then move them apart until, when touched, the person truly feels two points of contact. To ensure that the subject is feeling two points of stimulation, randomly touch the subject with one compass or divider point throughout this exercise. Record the distance between divider tips in millimeters or centimeters as measured with a ruler.

3. Repeat this procedure with the back of the hand, the lips or cheek, and the upper back or neck. (Each sense organ has finite dimensions and, therefore, only receives information from a discrete area of the skin. If receptors vary in density throughout the body, then the distance the divider points must be apart before two different receptors are touched also will vary throughout the body.)

4. Based on the distances found, which parts of the body are most sensitive?

5. On a piece of graph paper, draw circles of the diameters that you determined to be necessary for two-point discrimination at different areas of the body. This will represent graphically the differences in sensitivity between different parts of the body.

QUESTION: Based on the results just obtained, predict what the distances between receptors on other parts of the body might be. What about the toes, the soles, the legs, and the buttocks?

Observations:

Conclusion:

38 SECTION 3 The Integument

Hot, Cold, and Touch Receptors

Touch receptors, as well as hot and cold receptors, also are located in the dermal layer of the skin. They can be identified through use of hot, cold, and thermoneutral probes.

Hot, Cold, and Touch Receptors

Objectives:

- To appreciate the methods involved in mapping receptors in the skin.
- To understand punctate distribution of receptors.
- To realize that receptors for hot, cold, and touch sensations are distinct.

Materials:

- Small rulers.
- Black, blue, and red markers (permanent or water-resistant).
- Temperature probes.
- Hot-water source (hot plate, water bath, or hot running water).
- Cold-water source (ice or chilled water).
- Wooden applicator sticks.

Procedure:

1. Select a 2-cm square area on the inside of the forearm (one with as little hair as possible and not over any large superficial veins). Mark the area with a black marker (Figure 3.3). Your challenge now is to find all of the touch, hot, and cold receptors in the selected area.

2. For touch receptors, use a wooden stick to gently poke the entire area, moving several millimeters between poke (Figure 3.4). This is a blind test, so the subject should have his or her eyes closed or be looking away from the test area. Wooden sticks should be neither hot nor cold to the touch, thus not stimulating either hot or cold receptors.

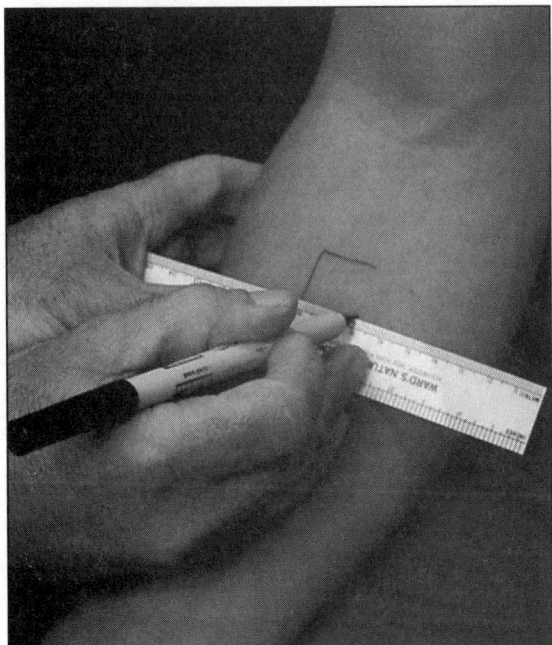

Figure 3.3
Selecting Test Area on the Skin.
Select a relatively hairless area on the inside of the forearm as the test area. It also is helpful to avoid large, superficial veins in the area. Draw a 2-cm square as the test area.

Figure 3.4
Testing for Touch Receptors.
Using a wooden stick as a probe, touch the entire test area, moving a few millimeters for each touch. Mark the spots that elicit a strong touch sensation with a black marker to indicate each touch receptor identified.

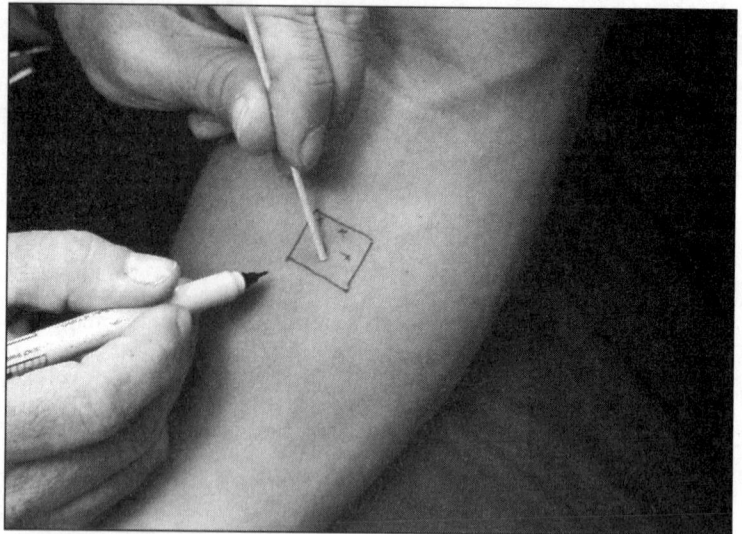

After a few pokes, the subject should be able to discriminate the slight sensation of some object on the skin from the distinct sensation of the wooden stick stimulating a touch receptor. Mark each such spot with a black marker. After some practice at definitely feeling a sensation from a touch receptor, it may be necessary to repeat some of the earlier spots to make sure these were true touch receptors.

3. There are two types of probes that can be used for detecting temperature receptors. First, blunt-metal dissecting probes that have been immersed in either hot (65°C) or cold (ice) baths can be used. Make sure the probe is dry, however, before touching it to the skin. The trouble with this type of probe is that it instantly and continuously loses or gains heat. Therefore, several probes should be kept in the baths so that a new warmer or colder probe can be used partway through the test. A second type of probe can be made from a 15-ml plastic syringe body (Figure 3.5). Remove the plunger and plunger shaft from the syringe barrel, then remove the rubber tip from the plunger. Insert copper wire of the proper gauge into the

Figure 3.5
Hot\Cold Probe.
A probe to test for hot and cold receptors can be made from a modified syringe barrel.

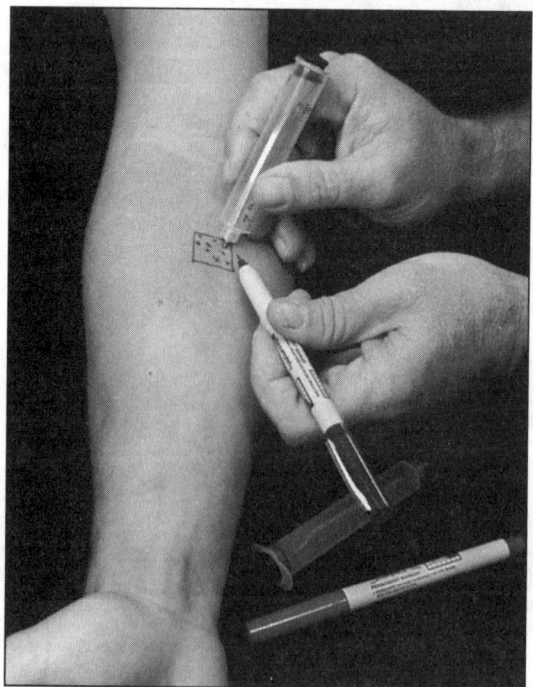

40 SECTION 3 The Integument

small opening of the syringe so that the wire is sealed in place and is long enough to run the length of the barrel. Fill the barrel with either hot or cold (ice) water, and cap with the rubber tip removed from the plunger. The copper wire is the probe tip and, depending on how it was cut, may have to be sanded or filed down to make it smooth. This probe will maintain its temperature longer and provide better discrimination of hot and cold receptors.

4. Continue testing the area, but this time, use the hot probes and then the cold probes. Mark the hot receptors with red markers and the cold receptors with blue markers.

Note: The pattern of the receptors you have just marked with black, red, and blue typically is not in some predictable, geometric configuration. This somewhat randomly distributed array of receptors is referred to as punctate distribution, which describes a similarity to the random arrangement of punctuation marks on a page of print.

5. In the box below, record the pattern of receptors found. Use the same pen colors as used to mark the skin.

QUESTION: Does the relative number of hot versus cold receptors found on you or your partner correlate with perception of hot and cold temperatures? In other words, does the person with lots of cold receptors report sensitivity to cold and the opposite with hot receptors? Why do you think this might be?

Observations:

Conclusion:

Triple Response

At one time or another, everyone has accidentally scratched themselves on a sharp object, resulting in a painful, swollen, red line or area on the skin. These changes on injury are referred to as the triple response, which represents the early stages of the inflammatory response. **Inflammation** is the response to injury that occurs as the body tries to limit and repair the resultant damage.

Triple Response

Objectives:

- To demonstrate the steps in the triple response.
- To appreciate the process of inflammation.

Materials:

- Alcohol pads.
- Sterile cotton-tipped applicator (wooden shaft).

Procedure:

1. This process can be simulated safely with a few precautions. Clean the inside of the forearm with an alcohol pad, then let the skin dry thoroughly.

2. Obtain a sterile wooden stick, such as the wooden-stick end of a sterile cotton-tipped applicator, and scratch the cleaned area on the forearm with the sterile stick. Do not break the skin, however. Apply just enough pressure to leave a white line in the epidermal layer of the skin. Remember that the outermost layer of the skin is the thick stratum corneum, which is composed of dead cells. The white line results from the disrupted corneal-cell layer.

3. If the skin is adequately stimulated, there is a consistent pattern of changes associated with the triple response: (1) An area parallel to the scratch line will become pale or white, which is caused by a local, transient vasoconstriction. Blood vessels leading into the area close up or constrict. Because skin color results not only from skin pigmentation but also from the dermal blood vessels, the skin becomes pale when the blood vessels close. This vasoconstriction or pale area may spread laterally from the "scratch." (2) Following the vasoconstriction is a longer-term vasodilation and increase in capillary permeability. Therefore, the scratch line becomes red, and this redness persists. (3) Because of the increased capillary permeability, there is an area of swelling or wheal where the skin was scratched. Hopefully, you did not scratch yourself this hard, but with an actual injury, there usually is pain associated with the process. Chemicals that cause pain are released from the damaged tissues.

QUESTION: The demonstration of triple response just performed was carried out on the skin. What do you think would happen with a similar injury inside the body? In other words, are different physiological processes at work in the skin compared with the rest of the body?

Observations:

Conclusion:

Rate of Nail Growth

The anatomy of fingernails and toenails includes the free edge, the nail body, and the nail root (Figure 3.6). The nail grows at the root and is constantly pushed out, so it must be cut at the free edge to maintain its length. Nail growth, like hair growth, is affected by nutrition, health, and age, and it varies from individual to individual.

Rate of Nail Growth

Objectives:

- To observe the growth of epidermally derived structures.
- To record quantitatively and consistently the growth of a fingernail.
- To carry out some simple comparisons to detect trends and relationships.

Materials:

- Volunteers.
- Sharp object or permanent marker.
- Ruler.
- Lab notebook.

Procedure:

1. Select a fingernail on each hand. Make a scratch in the nail with a sharp, pointed object toward the proximal end of the nail body, or mark the nail with a permanent lab marker. Do not cut across the nail body with a knife in a sawing action, because you may cut through the nail. Scratch a line several times with the sharp, pointed object until it is visible (Figure 3.7). It may be necessary to re-mark the scratch in the nail every week as it wears away through normal wear and tear on the nail body, or to re-mark with the lab marker as the mark washes off.

2. Over the next month, make daily or every-other-day measurements with a ruler of the distance from the scratch to the cuticle. Record this value in your lab notebook, and through your continuing measurements, determine the rate of growth for the nail. Compare your rate of growth with those of others in the class. Compare the rates for males and females. Is there any difference? Was the rate of growth consistent throughout the month?

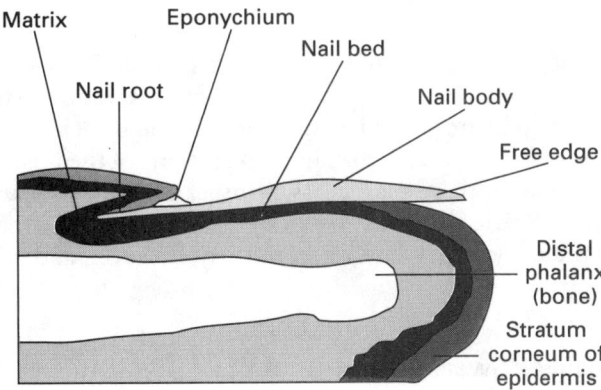

Figure 3.6
Section of a Fingernail.
This illustration shows the free edge, nail body, and nail root. Note the overlaying flap of skin (i.e., the cuticle). Remember that nails are epidermal in origin but extend down into the dermis.

Figure 3.7
Scratching the Fingernail.
To monitor the growth of a nail, a mark must be made in the nail body. Use a sharp object to scratch—not cut or saw—a mark in the nail, which can be followed over time to determine the rate of growth of the nail.

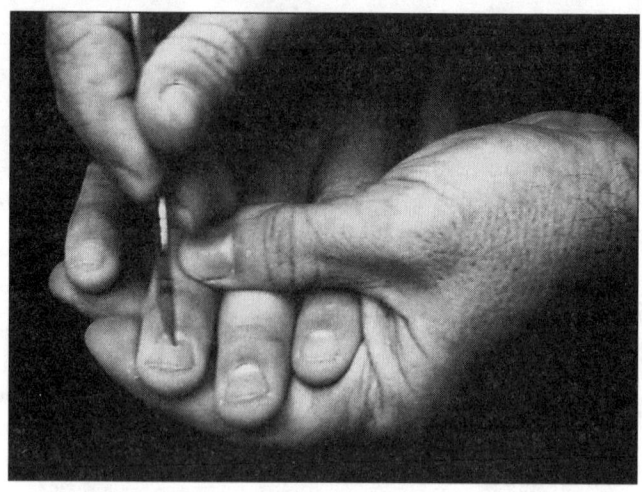

QUESTION: Hair is the other epidermally derived structure most of us have that also could be measured to determine its rate of growth. How would you measure hair growth? Do all hairs on the body grow at the same rate?

Observations:

Conclusion:

Demonstration of the Sweat Glands

Sweat glands eliminate waste materials from the body, and they also cool the body as sweat secreted on the surface of the skin evaporates. There are sweat glands over most of the body except under the fingernails and toenails, and on the glans penis and clitoris, labia minora, eardrums, and margins of the lips. In areas of the body where they are located, there may be as many as 3000 per square-inch (as on the palm of the hand). Sweat glands can be located and counted to determine their density at any part of the body. Hairs may get in the way, however, and might need to be removed to get a reliable count.

Counting Sweat Glands
Objectives:
- To learn how to demonstrate and count sweat glands.
- To appreciate the function of sweat glands as helpers in thermoregulation of the body.

Materials:

- Lab markers.
- Iodine solution, 1% iodine in 95% ethyl alcohol.
- Cotton-tipped applicators.
- Bond typing paper.
- Rubber gloves.
- Lab notebook.

Procedure:

Caution: If you are sensitive to iodine, do not volunteer to be a subject in this exercise.

1. Mark off a 4-cm² area on the palm of the hand or other relatively hairless area. Using a cotton-tipped applicator, paint the area with a 1% iodine solution in 95% alcohol, then let the solution dry completely.

2. While wearing rubber gloves, hold a 4-cm² piece of bond paper over the painted area for 30 seconds. Remove the paper, and note the blue dots, each of which each indicates the pore of a single sweat gland (Figure 3.8). Bond paper is required, because it contains starch, which will turn blue when sweat carries iodine onto the paper.

3. Repeat the procedure on another relatively hairless area of the body, and record the sweat gland count.

4. Other variations include repeating the procedure after light, and then heavy, exercise, or after going into a colder or warmer environment.

Figure 3.8
Counting Sweat Glands.
To count the sweat glands, place a piece of bond paper over the iodine-painted area for 30 seconds, then remove the paper. Every blue dot indicates the location of a sweat gland.

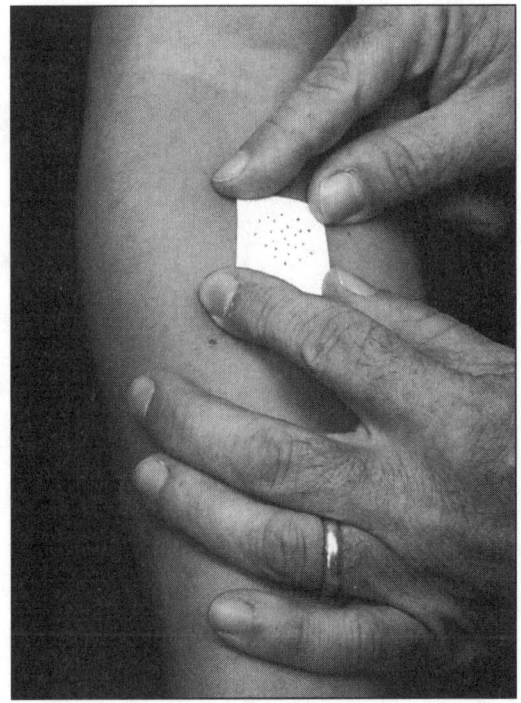

QUESTION: What do these observations tell you about the function of sweat glands? Do you think there is any correlation between the density of sweat glands and where a person grows up? In other words, do people who grow up in colder climates have fewer sweat glands than people in warmer climates?

Observations:

Conclusion:

Section 4

Bone and the Skeletal System

Note to the Student

As you prepare to study the bones of the human skeleton, you first must consider the internal structure of bones. Bones, and even the same bone, may have two different arrangements of bony material inside. A typical long bone, which often is described simply as a typical bone, is long, narrow, and has a cylindrical shape. There is a knob-like structure at each end, with a long shaft in between (Figure 4.1). The ends of the bone are referred to as the **proximal** and the **distal epiphyses** (singular, *epiphysis*) (*epi* (G), upon; *physi* (G), bladder, referring to the enlarged end of the bone). The long shaft is called the **diaphysis** (*dia* (G), apart, separate). The shaft is hollow, with a **marrow cavity** inside; the wall of the shaft is composed of dense or compact bone. Because this type of bony material is so dense, a special arrangement of tubes and channels is needed internally to nourish the living bone cells. These systems are called **Haversian systems** or **osteons.** The ends, or epiphyses, of the long bones are composed spongy or cancellous bone, which contains a mesh-like arrangement of bony struts called **trabeculae** (*trabecula* (NL), marked with crossbars). The spaces around and between these supports are filled with red marrow like the marrow cavity. Because the bony supports are thin and the marrow is so rich in blood vessels, no special arrangement, such as that found in compact bone is needed to nourish the bone cells here.

Journal Note Have you ever broken a bone? If so, were any problems associated with its healing? Was that part of your body put in a cast, or was the bone reinforced with plates, screws, or both? Does the bone ever bother you, especially when the weather changes? Consider that a broken nose often leads to breathing problems later in life, because the spaces of the nose are modified. For example, snoring often is a problem after a broken nose.

Figure 4.1
A Typical Long Bone.
This typical long bone shows the parts associated with this shape of bone, including the epiphyses, diaphysis or shaft, epiphyseal plate, and marrow cavity. The ends or epiphyses typically are spongy or cancellous bone, and the walls of the shaft are compact or dense bone.

A – Proximal epiphysis
B – Diaphysis
C – Distal epiphysis

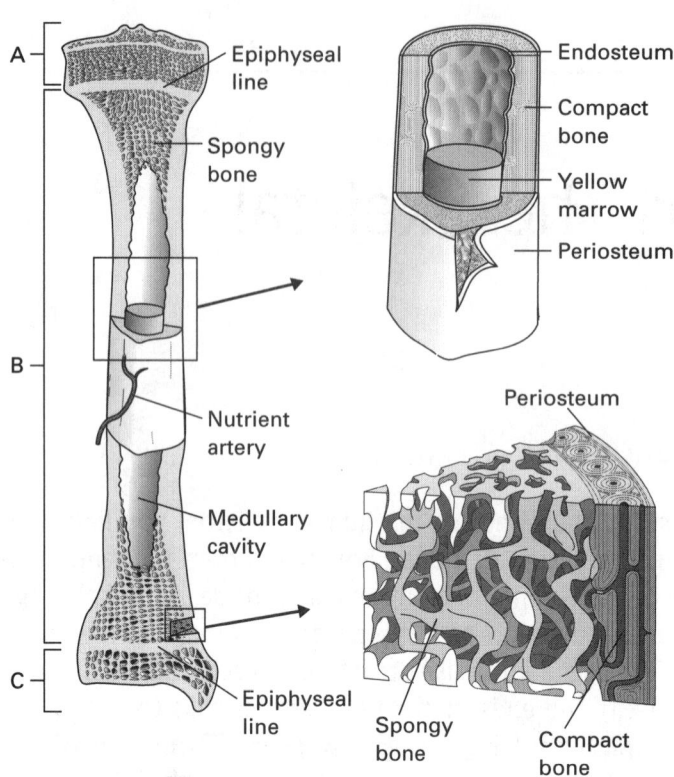

The Histology of Compact Bone

The classic slide of bone is a ground bone slide. As you observe this preparation, note that the tissue is not stained; the light and dark areas are a characteristic of the preparation. Find a typical Haversian system or osteon. Look for the central, dark **Haversian canal** that contained blood vessels and nerves in the living tissue. Surrounding the central canal are layers of hard, bony matrix, or **lamellae.** In between the lamellae are rows of bone cells called **osteocytes** (*osteo* (G), bone), which appear in spaces left in the bony matrix called **lacunae.** As bone initially forms, it is populated by cells called **osteoblasts** (*blast* (G), a bud, a sprout). These are the cells that actively secrete the bony matrix. They also surround themselves with bone and, when they no longer have any place to put more bone, are referred to as osteocytes. To stay alive so they can nurture the bony matrix, these cells maintain extensions between lamellae, through spaces that look like cracks in the bony matrix, which are called **canaliculi.** Osteons run the length of long bones; however, there also are canals that run across the long dimension of these bones. These are **perforating** or **Volkmann's canals.** Blood vessels and nerves enter the bones through these are channels and supply all the Haversian systems and the marrow cavity. Another element of connective tissue associated with bone is an outer covering, the **periosteum.** The inner marrow cavity in long bones is lined with its own connective tissue layer, the **endosteum.** Both of these layers have bone-building cells, the osteoblasts, and bone-destroying cells, the **osteoclasts,** associated with them. Bone is a living, growing tissue that changes as needed to meet life's demands.

Journal Note Observing the Meat We Eat. If you like to eat meat, observe the internal structure of bone in the cuts of beef and pork that you prepare at home or order in a restaurant. Does the bone look like a long bone? Is it compact or spongy bone? Also, note the marrow cavity. Does the marrow look like red or yellow marrow? (Most of the body's red marrow in the long bones becomes fatty, yellow marrow as we mature.) What does that tell you about the age of the animal from which the meat came? Chicken bones are quite different in size, but observe these same details about these bones as well.

The Histology of Compact Bone

Objectives:

- To understand the internal structure of compact bone.
- To realize the physiological function of the Haversian system in the maintenance of compact bone.
- To appreciate different techniques of slide preparation.

Materials:

- Microscope.
- Slide of bone, ground.

Procedure:

1. Place a slide of ground bone on the microscope stage, and carefully bring it into focus on low power. (These slides are labelled as "ground" bone, because bone cannot be sliced in the same fashion as other tissue to be made into sections for slides. The bone must be ground until it is thin enough to allow light to pass through. Be careful with these slides. They are thick sections, and care must be exercised at powers greater than ×100.)

Caution: When looking at slides of ground bone, exercise care so as not to damage the section.

---

2. Identify all of the structures mentioned in the previous discussion regarding the histology of compact bone and as illustrated in Figure 4.2.

QUESTION: When a bone is broken, damage occurs to all of its components. Based on the previous description of bone histology, speculate as to the process of bone repair.

Observations:

Conclusion:

Figure 4.2
A Section of Compact Bone. A section of bone is referred to as ground bone; Haversian systems or osteons are found in a cross-section of bone. Haversian systems are necessary to supply nutrition to the bone cells.
(a) Drawing of a section of compact bone.
(b) Photomicrograph of ground bone.

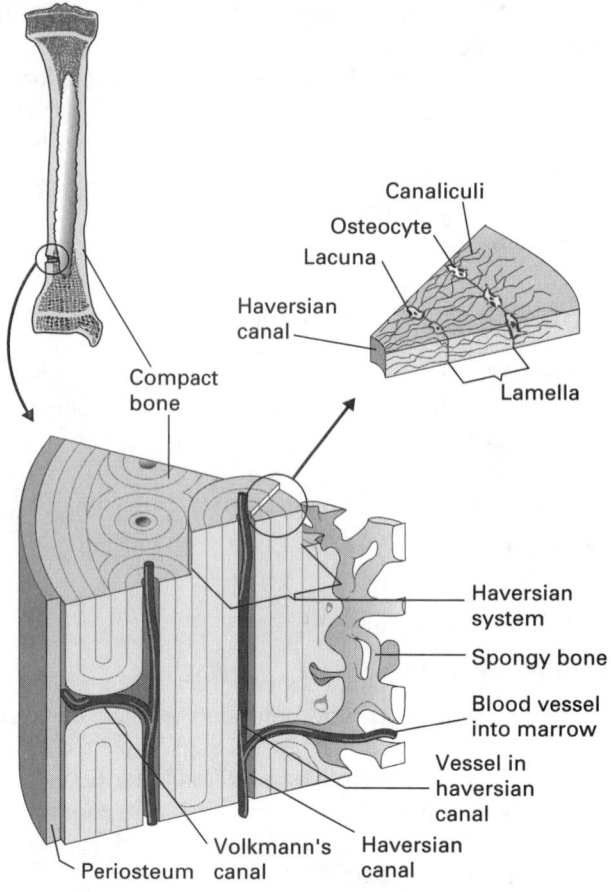

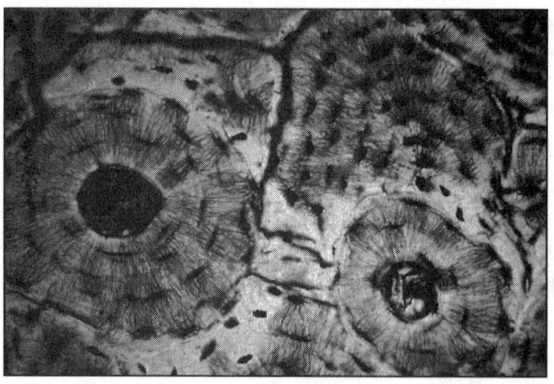

The Skeleton

The typical human skeleton is composed of 206 bones. This number may sound overwhelming, but realize there are two arms and legs and associated supporting structures. Once the bones of either arm or leg are known, the bones of the other are as well. Also, the terminology of bones or bony structures learned at this point will be useful throughout the course, because many of these terms also are the names for many other components of the body, such as the blood vessels, muscles, and nerves.

If your class is fortunate enough to have real bone material, here are a few suggestions to help preserve the bones for future students to use:

1. Wash your hands before working with the bones, because body grease and food remnants will darken and obscure details.

2. It is only natural to point at structures, sutures, and so on as part of the learning process. Without thinking, however, students sometimes will do so with pens and pencils, which will leave marks on the bones. Wooden sticks should be provided for use as pointers.

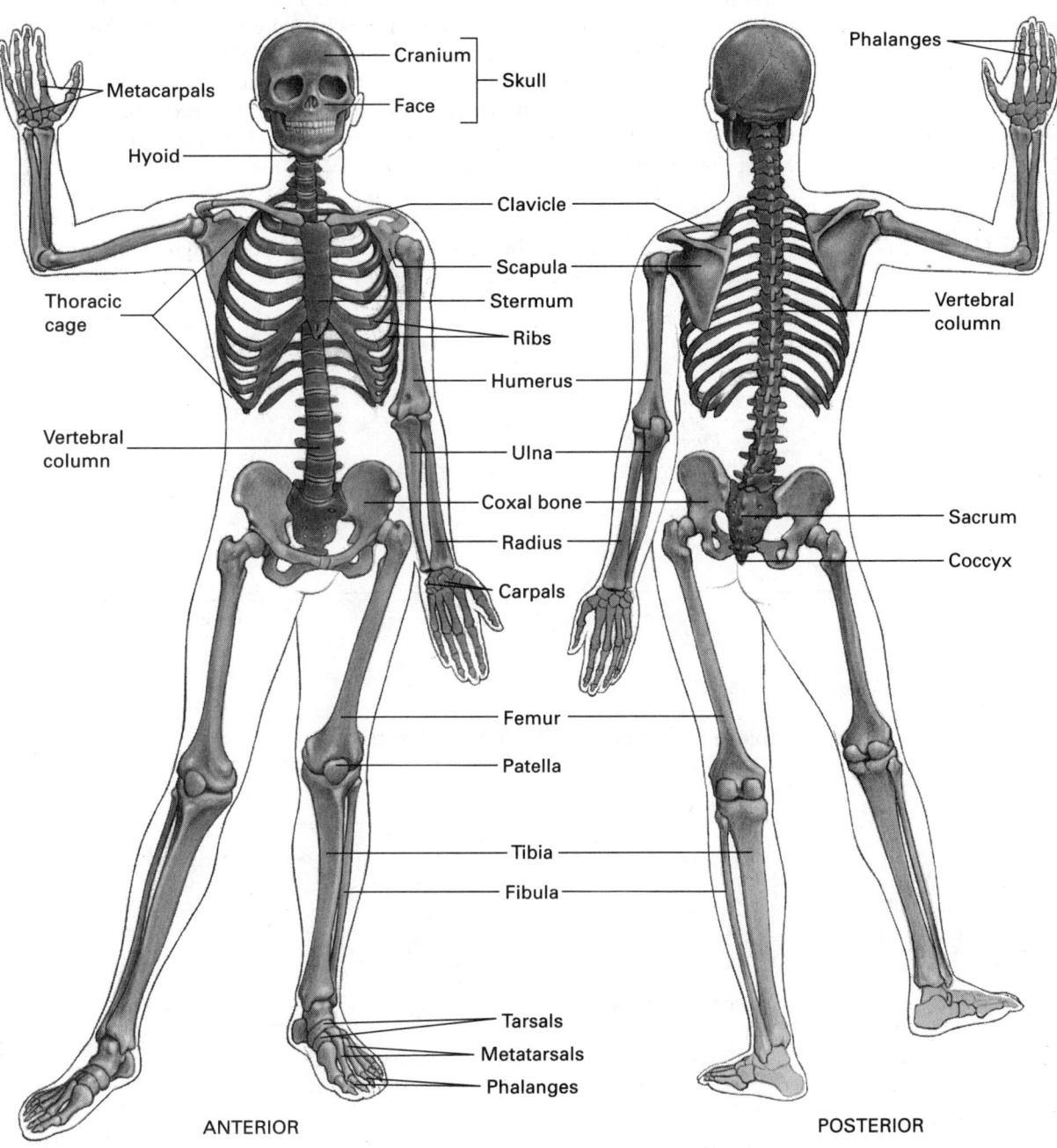

ANTERIOR POSTERIOR

Figure 4.3
The Human Skeleton.
The skeleton is composed of two parts: the axial skeleton, which is composed of the skull, vertebral column, and rib cage; and the appendicular skeleton, which is composed of the limbs (i.e., arms and legs) and their associated attachment structures (i.e., the pectoral and pelvic

The Skeleton

3. Table tops are hard surfaces. Padded paper cushions can be bought in rolls and cut in squares to place on the lab bench to protect the bones. Check to see if your instructor has such padding available.

4. Whenever you carry bones from one location to another, exercise special care not to drop them.

The skeleton is divided into two major portions. The **axial skeleton** is composed of the bones of the skull, vertebral column, and rib cage. These bones are positioned around the central axis of the body (Figure 4.3). The **appendicular skeleton** is composed of the arm and leg bones and their associated girdles or attachment structures (Figure 4.3). Tables 4.1 and 4.2 list the bones in each of these divisions. This section also contains figures showing the bones of the skeleton and describing the structures found on each. Rather than describe the structures in the text and then refer to the same items in the figures, the bones are identified by labels that contain information about the structure in parentheses.

QUESTION: What happens to the skeleton of a person who begins a bodybuilding program? Is it only the muscles that get bigger with weight-lifting?

Observations:

Conclusion:

Table 4.1 Bones of the Axial Skeleton

Skull

1 Frontal	2 Parietals	1 Occipital
2 Temporals	1 Sphenoid	1 Ethmoid
2 Zygomatics	2 Maxilla	1 Mandible
2 Palatines	1 Vomer	2 Lacrimals
2 Nasals	1 Hyoid	

Vertebral Column

7 Cervical vertebrae	12 Thoracic vertebrae
5 Lumbar vertebrae	5 Sacral vertebrae
4 Coccygeal vertebrae	

Rib cage

12 Thoracic ribs	Manubrium sternum
Body of the sternum	Xiphisternum

Table 4-2 Bones of the Appendicular Skeleton

Each Upper Limb (Arm)

1 Scapula	1 Clavicle	
1 Humerus	1 Radius	1 Ulna
8 Carpals		
1 Hamate	1 Pisiform	1 Triquetral	1 Lunate
1 Capitate	1 Scaphoid	1 Trapezium	1 Trapezoid
5 Metacarpals	5 Proximal phalanges	
4 Middle phalanges	5 Distal phalanges	

Each Lower Limb (Leg)

1 Innominate or coxal (ischium, ilium, and pubis)		
1 Femur	1 Patella	1 Tibia
1 Fibula		
7 Tarsals		
1 Calcaneus talus	1 Navicular	1 Cuboid
1 Medial cuneiform	1 Intermediate cuneiform	
1 Lateral cuneiform		
5 Metatarsals	5 Proximal phalanges	
4 Middle phalanges	5 Distal phalanges	

Journal Note Occupations and Body Shape. Have you ever noticed the shape of people who work or perform in a physically demanding occupation or sport? Note the shoulders of swimmers, and imagine what the bones of the upper body, the arms, and their associated girdles look like. Wrestlers tend to be short and stocky. What other occupations or activities are associated with specific body shapes or dimensions? Do your own specific activities have effects on the structure of your body?

Learning the Bones of the Axial Skeleton

Objectives:

- To appreciate the bones that make up the axis of the body.
- To learn the bones of the axial skeleton so their names will be useful in learning others parts of our anatomy, such as the muscles and blood vessels.

Materials:

- Human skeletal material, real bone or plastic.

The Skull

Like much of the axial skeleton, the skull is characterized by many of its bones being present as one rather than as a pair (Figure 4.4). The bones of the skull are divided into **cranial** and **facial bones,** a distinction that arises because some bones of the skull do not make contact with the brain (i.e., the facial bones). The cranial bones include the frontal, the two parietal, the occipital, the two temporal, the sphenoid, and the ethmoid bones, for eight bones in all. The facial bones are the remaining bones of the skull, for a total of 14. The names of the larger, superior bones of the skull are the same as the lobes of the brain.

Figure 4.4a provides an anterior view of the skull, and Figure 4.4b provides a lateral view. Both views are required to understand the position, dimensions, and connections between the bones of the skull. Use these two illustrations along with the following discussion to learn the names of these bones.

Figure 4.4
The Skull.
(a) This frontal view shows bones considered to be part of the cranium (i.e., cranial bones) and to be part of the face (i.e., facial bones). (b) This lateral view shows the extent of the cranial bones that complete the braincase.

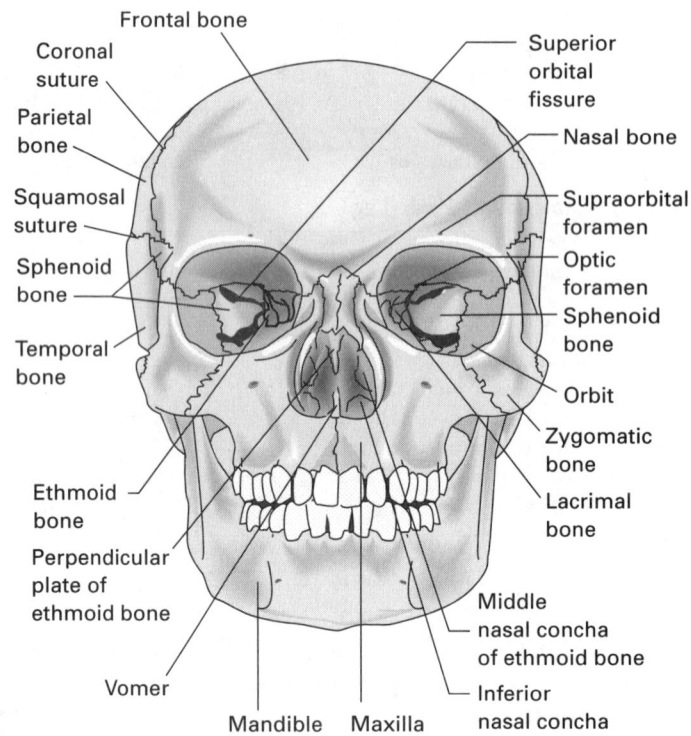

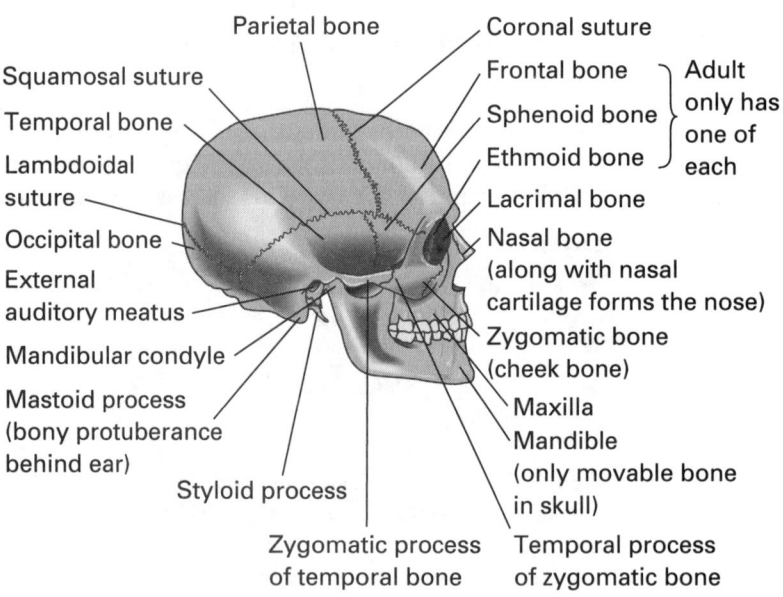

54   SECTION 4   Bone and the Skeletal System

The Cranial Bones

The Frontal Bone

The frontal bone makes up the anterior of the skull, and it forms sutures or seams with the parietal bones, the sphenoid and the ethmoid, the two zygomatic, the two lacrimal, the two maxilla, and the two nasal bones. A **suture** is a type of immovable joint where two bones overlap or interdigitate. Sutures either have specific names or are named based on the two bones that connect at that point. The suture between the frontal and the parietal bones is called the **coronal suture.** Its other sutures are named based on the bones connected together. The frontal bone forms the arch that makes up the top part of the opening for the eye, or the **orbit.** The top edge of the orbit is the **supraorbital margin,** and it contains either a foramen or a notch for blood vessels and nerves supplying the face, which is called the **supraorbital foramen** or **notch.** Note the functionality of the names for this structure. Does the skull you are observing have a supraorbital notch or foramen?

The Parietal Bones

The two parietal bones form the **sagittal suture** along the midline at the top of the skull. They also overlap the edge of the temporal bone and from there the **squamosal suture.** The **lambdoidal suture** is located where the parietals connect with the occipital bone. There sometimes are extra small bones associated with suture lines, and this is especially so of the lambdoidal suture. These are called the **Wormian** or **sutural bones.**

The Occipital Bone

The occipital bone "cups" the posterior aspect of the brain surrounding the occipital lobes and the cerebellum. The occipital bone contains the largest foramen in the skull, the **foramen magnum,** which is the point of entry for the spinal cord into the cranial cavity.

The Temporal Bones

There are two temporal bones in the skull, with each bone overlying the temporal lobe of the brain. These bones are placed laterally on the skull and are the only cranial bones not in contact. Each temporal bone contains an extension that articulates with the zygomatic bone on that side of the face. This extension is called the **zygomatic process of the temporal bone.** A comparable extension of the zygomatic bone also exists, an these two processes make up the "cheekbone," or **zygomatic arch.**

The Sphenoid and Ethmoid Bones

The sphenoid and the ethmoid are similar in several ways. They are singular bones, and, therefore, are located on the midline of the body. They not only make contact with the brain but also contain important structures associated with the special senses (the olfactory and visual senses) and the cranial nerves. The sphenoid, which is the larger of the two, has two laterally projecting, wing-like structures called the **greater** and the **lesser wings.** The **optic foramen,** which is the opening for the optic nerve (i.e., cranial nerve II), can be observed. If a sagittally sectioned skull is available, identify the **sphenoidal sinuses.** The **sella turcica** (*sella* (L), a saddle; *turcica* (NA) Turkish), which surrounds and protects the pituitary gland, is located superior to the sinuses.

QUESTION: Pituitary tumors cause a variety of symptoms because of the many hormones produced by this gland. Considering the "buried" location of the pituitary gland, how do you think pituitary tumors are removed surgically?

Observations:

Conclusion:

The ethmoid bone covers the nasal passageways and contains a perforated plate known as the **cribriform plate** (*cribr* (L), a sieve). These numerous perforations are for the olfactory nerves (i.e., cranial nerve I) to enter the cranium and connect to the olfactory bulb. A thin, bony, superior extension, the **crista galli** (*crista* (L), a crest; *gallus* (L), a chicken, cock), extends between the two frontal lobes. A midline, inferior-projecting plate, the **perpendicular plate,** forms part of the **nasal septum.** Other extensions of the ethmoid form the **nasal conchae** (*concha* (G), a shell) in the nasal passages as well as the wall separating the nasal passages and the orbit. (The pair of inferior nasal conchae are separate and are considered to be facial bones.) All of these bony projections are covered with mucous membrane, which helps the nasal passages do their job of cleaning, warming, and moistening the air that we breathe.

The Facial Bones

The Nasal Bones

The nose is composed of a pair of nasal bones and associated cartilages. All of us are familiar with a typical skull; in fact, you probably have a skull in front of you now. The reason why the nose always appears to be missing is because the cartilages associated with the nose are gone.

The Maxilla

The upper jaw is composed of a pair of maxilla. This is the rigid jaw as it articulates with many other bones of the face. Half of a person's teeth are located in the upper jaw, and they typically articulate with the nasal and frontal bones superiorly and with the zygomatic and palatine bones laterally and posteriorly. A plate-like, internal extension forms the anterior portion of the roof of the mouth. There usually is a hole, the infraorbital foramen, in the maxilla through which nerves and blood vessels make their way to the face.

The Palatine Bones

The palatine bones make up the rest of the roof of the mouth, or the hard palate, along with the maxilla.

The Zygomatic Bones

Along with the temporal bones, the zygomatic bones make up the cheeks. They form a large portion of the orbit that contains the eyes, and an extension of the zygomatic bone, called the temporal process, connects to the zygomatic process of the temporal bone to form the zygomatic arch.

The Lacrimal Bones

The lacrimal bones are very small, thin bones located between the ethmoid and the maxilla. There is a groove through this bone for the lacrimal duct to drain tears into the nasal passageways. (Have you ever noticed how your nose "runs" when you cry?) These bones are easily damaged, so be careful when handling your skull.

The Vomer

The vomer is one of the singular bones of the skull. It makes up a major portion of the nasal septum, which is the wall dividing the left and right nasal passageways.

The Mandible

The lower half of a person's set of teeth is located in the sockets of the mandible. The back edge of the mandible, or jawbone, is palpable at the edge of your face, slightly anterior from the ears. There is another obvious opening in the jawbone, the mental foramen, through which blood vessels and nerves pass to the face. The posterior-superior part of the mandible is called the ramus, and its rounded posterior process is called the condyloid process, which is where it articulates with the temporal bone. This forms the temporomandibular joint. The anterior projection of the ramus is the coronoid process, which is where the temporalis muscle connects to the mandible.

Journal Note A good exercise and study aid is to name all the bones that connect to a certain bone of the skull. For example, the frontal bone of the skull connects to 12 other bones. Can you name them?

The Rib Cage and Vertebral Column

The rib cage and vertebral column make up the rest of the axial skeleton. Figures 4.5 through 4.7 show the bones of these areas. Refer to these illustrations for details regarding structure.

Journal Note Surgery Involving Bones. The pituitary gland is difficult to reach surgically. What other parts of the body require that bone be removed, cut, or altered as part of the surgical procedure to reach them? What about the sinuses? The brain?

Learning the Bones of the Appendicular Skeleton

Objectives:

- To appreciate the bones making up the appendages of the body.
- To learn the bones of the appendicular skeleton so their names will be useful in learning others parts of our anatomy, such as the muscles and blood vessels.

Materials:

- Human skeletal material, real bone or plastic.

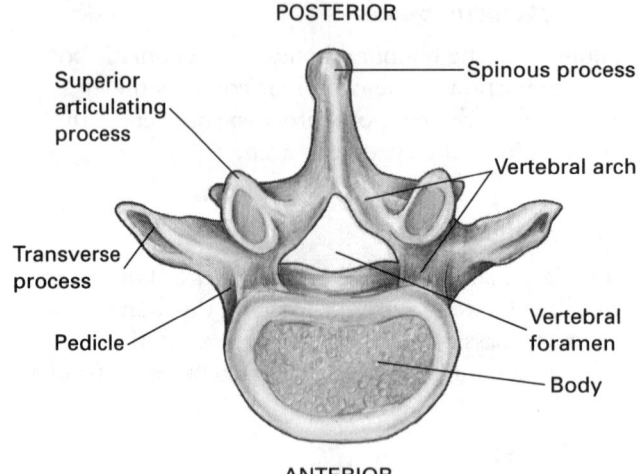

Figure 4.5
A Vertebra.
There are differences between vertebrae that make up different regions of the vertebral column. This illustrations shows the typical parts.

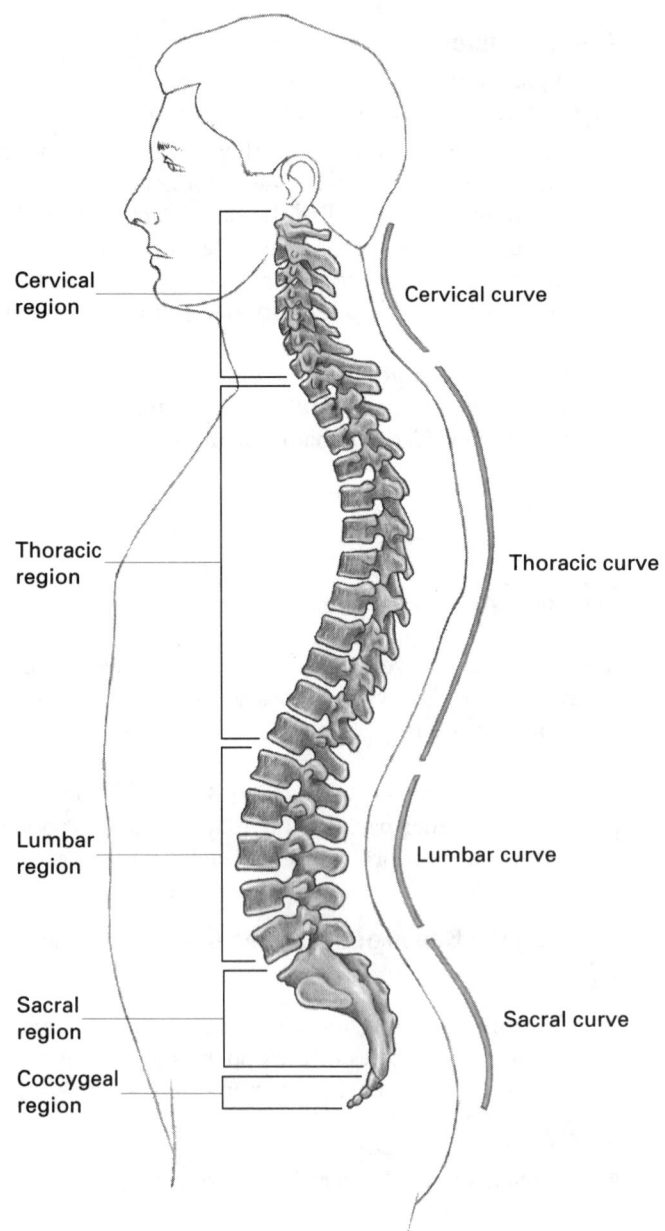

Figure 4.6
The Vertebral Column.
Typical curves to the adult vertebral column are shown.

Figure 4.7
The Rib Cage and Typical Rib.
The rib cage provides protection for many of the internal organs of the chest. It also supports the soft, spongy lungs.

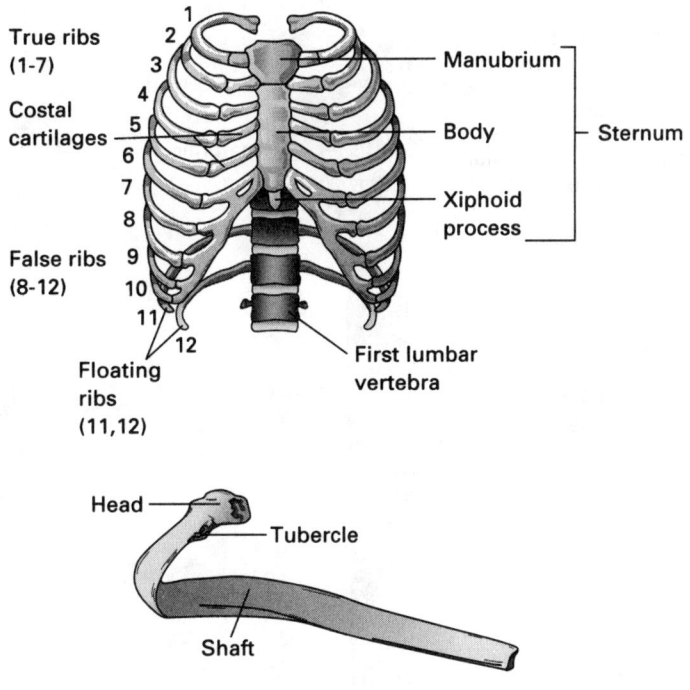

The Upper Limb (The Arm)

When discussing the limbs, there are two sets of components to study: the limb bones themselves, and the associated attachment structures. Figure 4.8 illustrates the scapula, which is the principal bone of the pectoral girdle (i.e., the attachment structure for the arm) Figures 4.9 through 4.11 show the bones of the upper limb or arm.

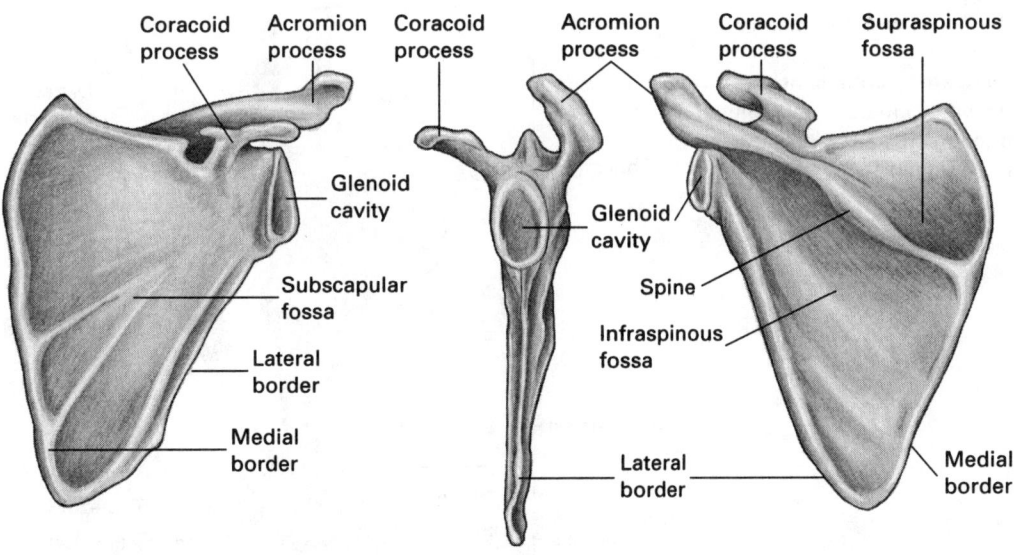

Figure 4.8
The Scapula.
The scapula is the principal bone of the pectoral girdle.

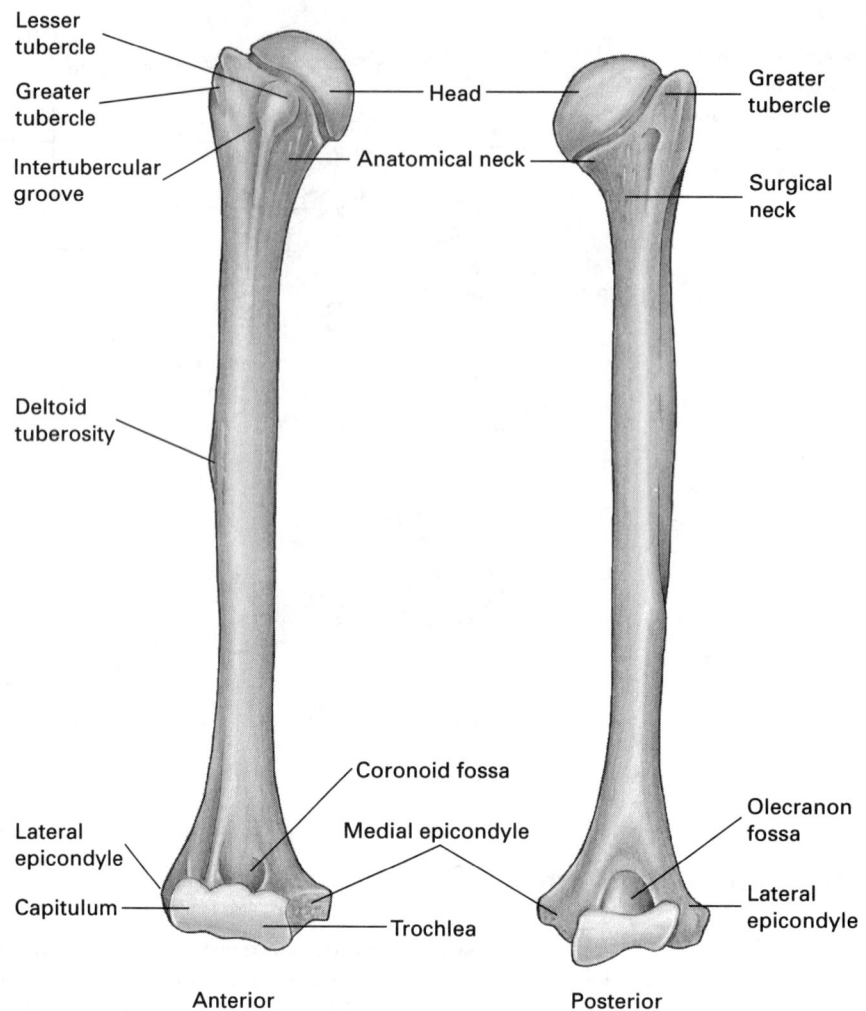

Figure 4.9
The Humerus.
The humerus articulates with the glenoid fossa, which is formed by the scapula. This is a much more flexible joint than the hip joint.

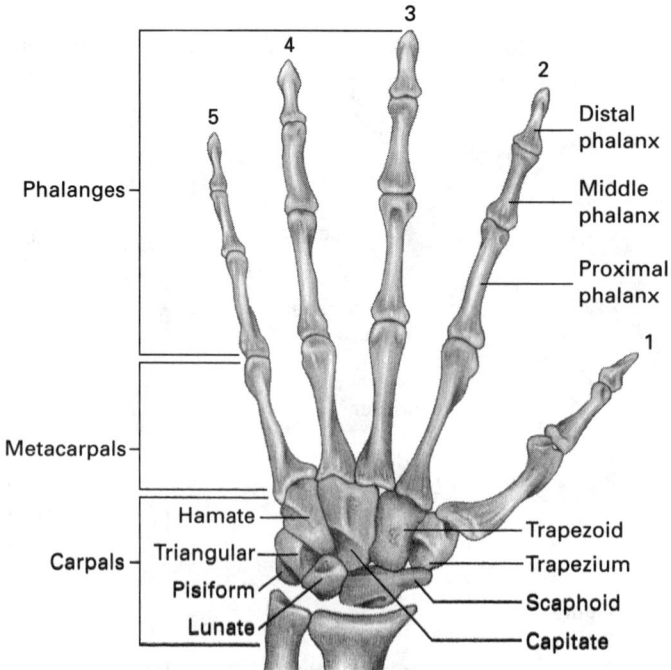

Figure 4.10
The Bones of the Wrist and Hand.
There are 8 carpal bones on the wrist, 5 metacarpal bones in the palm, and 14 phalanges in the fingers.

60 SECTION 4 Bone and the Skeletal System

Figure 4.11
The Ulna and Radius.

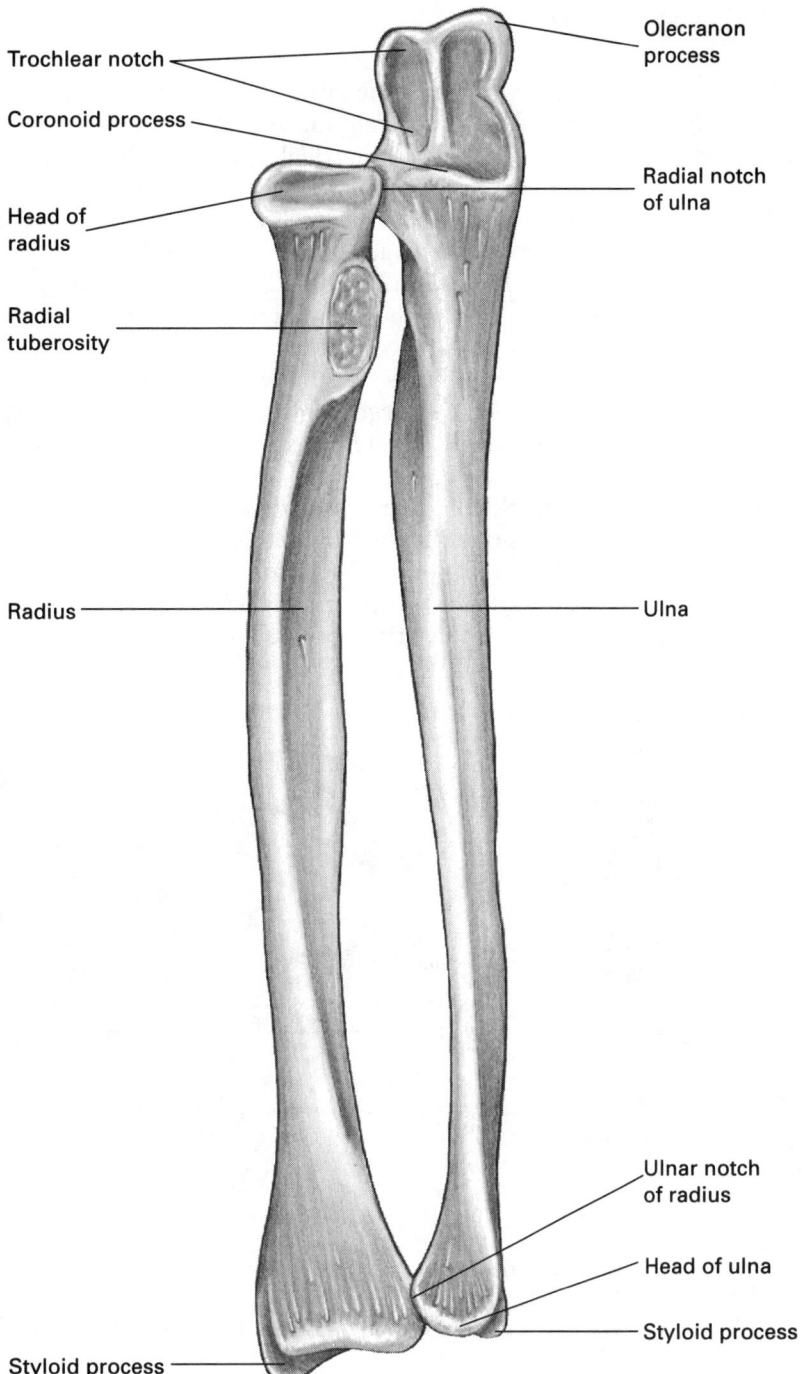

The Upper Limb (The Arm)

The Lower Limb (The Leg)

The study of the pelvic girdle, which is the attachment structure for the lower limbs (Figure 4.12), is different for several of reasons. For one, to discuss the girdle requires inclusion of the sacrum, which actually is the five fused sacral vertebrae that are part of the vertebral column, which in turn is part of the axial skeleton (Figure 4.13). Another difference is that the coxal bone really is the result of the fusion of three embryonically separate bones. Figure 4.14 shows the extent of the three separate bones of the pelvis: the ilium, the ischium, and the pubis. Note that part of each bone comprises part of the acetabulum, which is the socket that receives the head of the femur.

The lower limbs are similar in structure to the upper limbs. Figure 4.15 displays the femur, or the large thigh bone of the leg. Figure 4.16 illustrates the lower leg bones, or the tibia and fibula. Figure 4.17 shows the ankle and foot bones.

Journal Note Mnemonic (*mnem, -at, -on* (G), memory, remember) devices are lists or rhymes that help us to remember lists or important rules. One that I often quote is "Do as you oughta, add acid to watah" to remind me it is unsafe to pour water into acid. Can you come up with any devices or rhymes to help you remember parts of the body, such as the wrist or ankle bones?

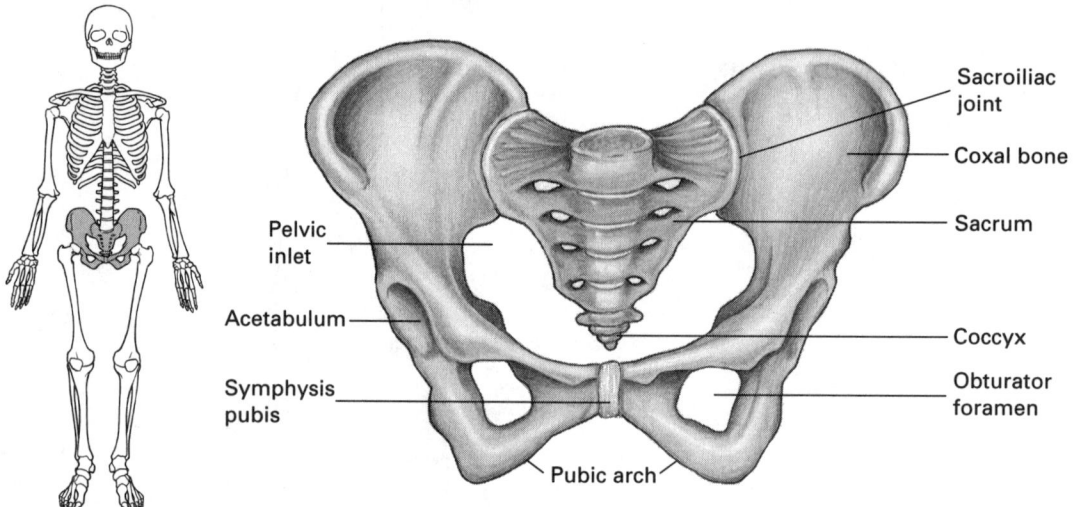

Figure 4.12
The Pelvic Girdle.
The pelvic girdle contains two pelvic bones and the sacrum.

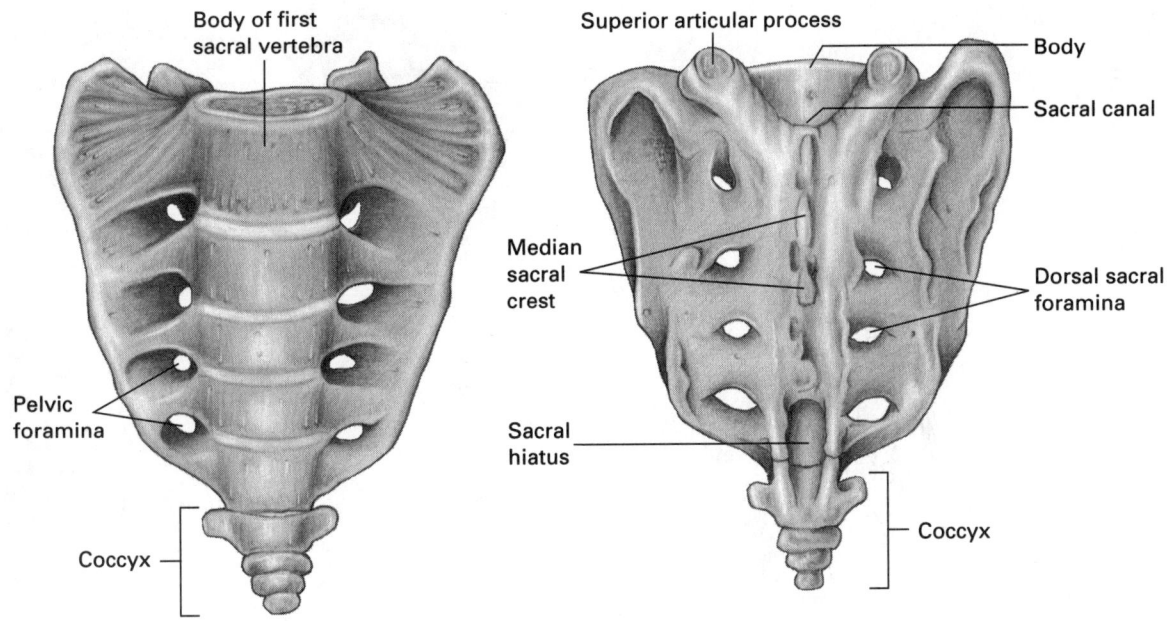

Figure 4.13
The Sacrum.
Fusion of the five sacral vertebrae produce a solid attachment structure that connects the lower limbs with the rest of the body.

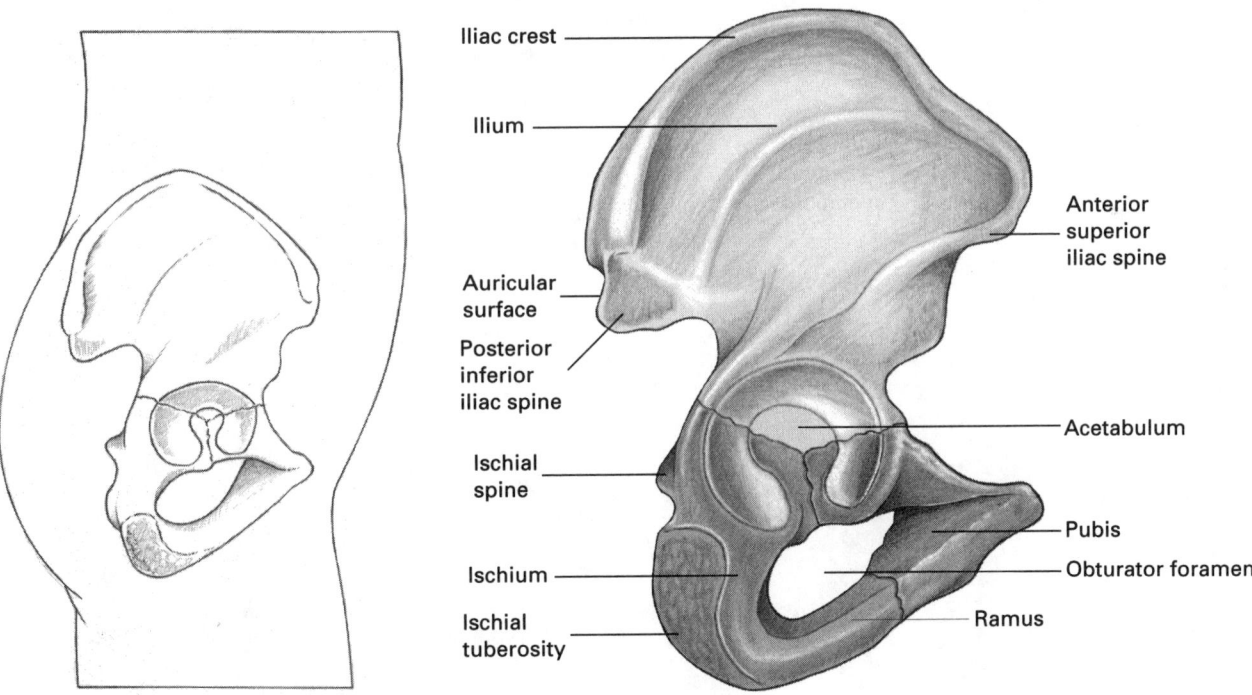

Figure 4.14
The Pelvic Bone.
There are several names for the pelvic bone, including the os coxae, innominate bone, and coxal bone.

The Lower Limb (The Leg)   **63**

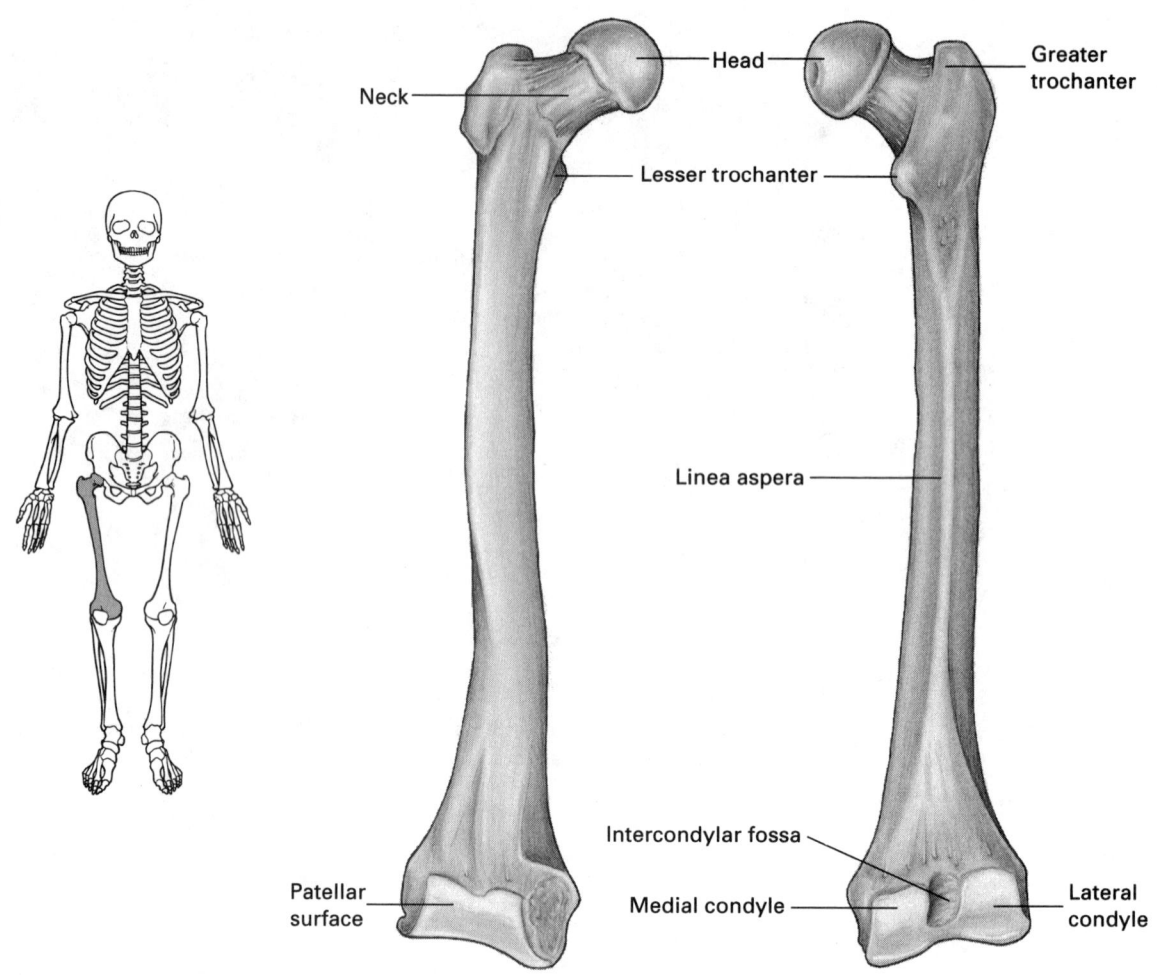

Figure 4.15
The Femur.
The femur is the longest and strongest bone of the body.

SECTION 4 Bone and the Skeletal System

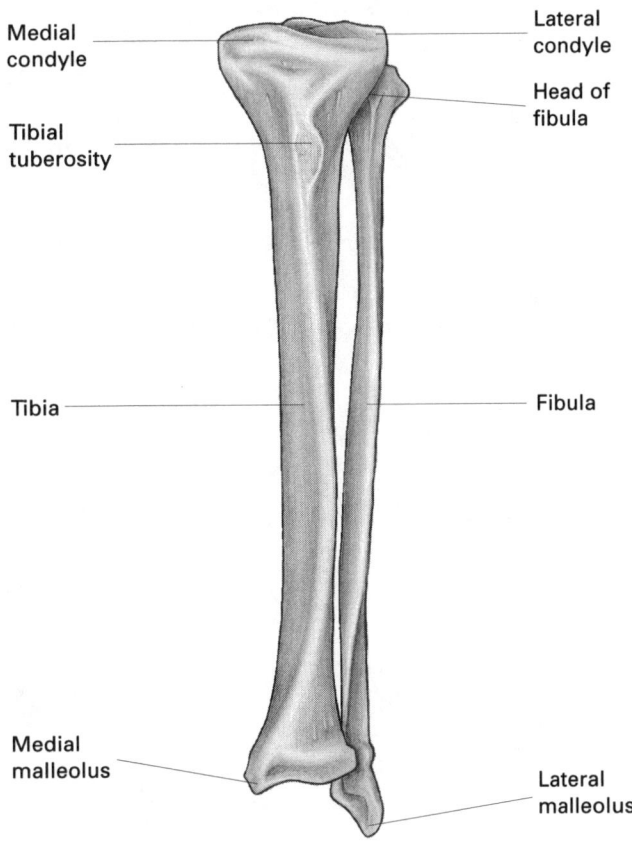

Figure 4.16
The Tibia and Fibula.

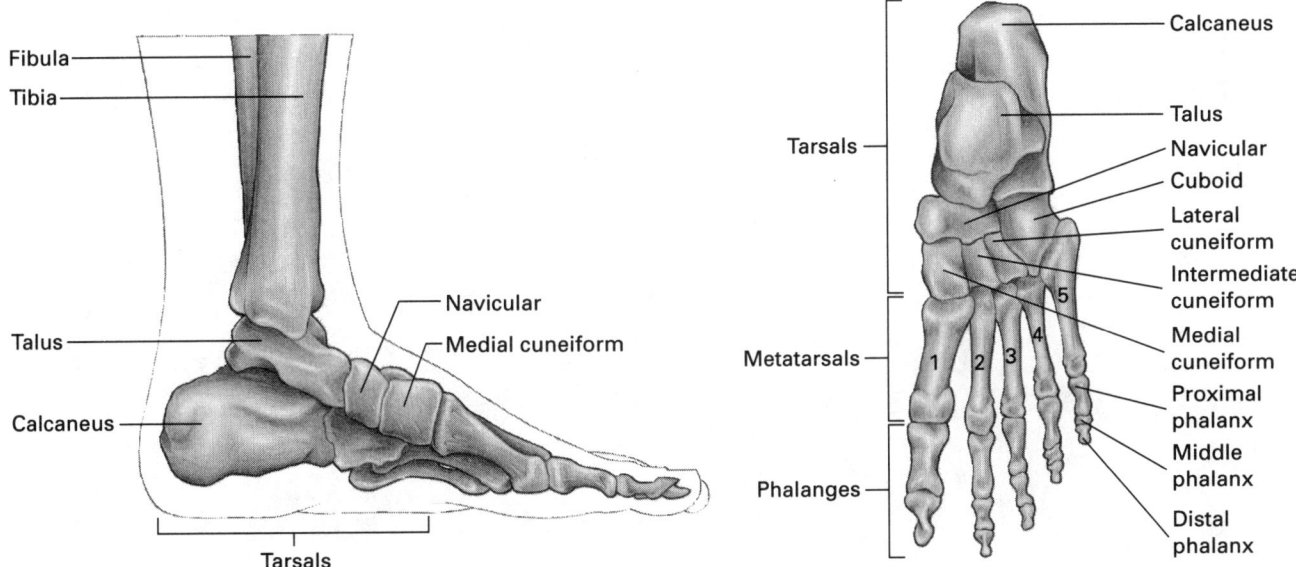

Figure 4.17
The Bones of the Ankle and Foot.
There are 7 tarsal bones in the ankle, 5 metatarsal bones in the arch of the foot, and 14 phalanges in the toes.

Section 5

The Muscular System

Note to the Student

The muscular system refers to the muscles that are attached to the skeleton, so we actually are referring to the skeletal muscles. Two other types of muscle (i.e., smooth and cardiac) will be mentioned when we discuss the systems they are most involved with: the cardiovascular, digestive, urinary, and reproductive systems.

Skeletal muscles are attached to bones by tendons. The bones serve as levers and the joints as fulcrums, thus allowing the muscles to provide a source of power or effort when moving parts of the body. Muscles are made of large cells, or the muscle cells, which also are known as muscle fibers. Much of an individual's total mass is comprised of the muscle in the arms and legs. The waste heat from muscular activity is used to maintain the body temperature in warm-blooded animals; therefore, muscles are important for homeostasis.

Muscle tissue has the following four properties:
1. Contractility: the ability to shorten in length.
2. Excitability: the ability to respond to a stimulus.
3. Extensibility: the ability to be stretched.
4. Elasticity: the ability to return to its original size.

The Histology of Muscle

The structural and functional unit of skeletal muscle is the **sarcomere** (Figure 5.1). This basic unit contains all the components associated with the activity of skeletal muscle. A subcellular structure, the sarcomere is composed of **thin** and **thick filaments**. Recall from the section on cell structure the description of microfilaments and microtubules; the thin and thick filaments mentioned here are examples of microfilaments.

The thin filaments actually are composed of many spherical **actin** molecules, which are strung together like beads on a string. It may be hard to conceive how bead-shaped molecules could make a thin filament, but to picture this, simply pull a strand of beads until taut, thus creating a linear arrangement (Figure 5.2).

Figure 5.1
The Sarcomere.
The sarcomere is the structural and functional unit of muscle tissue. It is composed of thin and thick filaments, which contain actin and myosin, respectively. The areas of the sarcomere are referred to as the A band and the H zone. The I band runs from one sarcomere to the next, and the Z lines are the lateral limits of each sarcomere. (a) A realistic representation of a highly magnified sarcomere. (b) A schematic representation of a sarcomere.

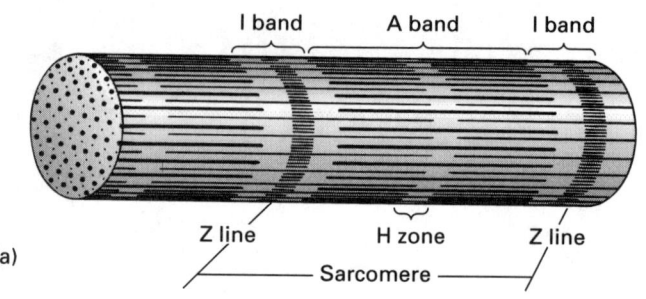

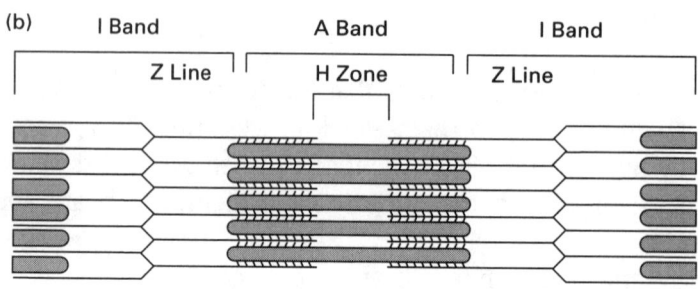

Figure 5.2
The Thin Filament.
One model of the thin filament molecular structure is a string of beads (a), which can be pulled tight so that the two rows run along each other (b). In actual thin filaments (c), there is a twist or spiral down the length of the chain. Note how the chain has been twisted so there are seven beads per complete turn in the molecule. Note also that other molecules besides actin are associated with the thin filament.

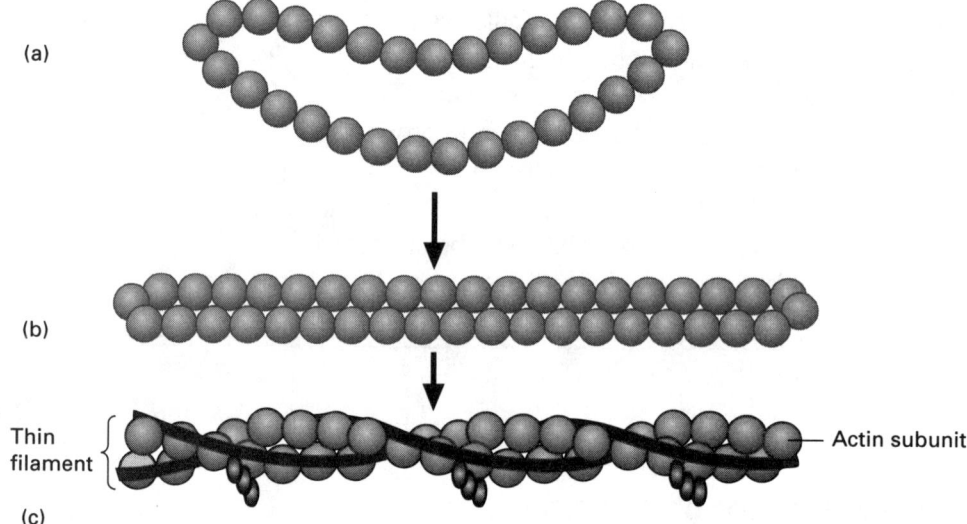

There is also a twist in the thin filament. Note that the beads on your strand can be twisted a little, so that there are quite a few beads per turn, or very tightly, so that there are very few beads per turn. In the actual thin filament, there are seven beads per turn.

The thick filaments are composed of golf club–shaped molecules, which are called the **myosin** molecules. This means they have a region that looks like a handle (tail) and a club portion (head). To assemble these molecules into a thick filament, a bundle of myosin molecules are combined much like you would gather a set of golf clubs into a bundle (Figure 5.3). Picture the group of golf clubs so that all of the club portions are sticking out from the handles. In a similar way, the thick filaments are long structures made of the aligned tails, with the myosin heads sticking out from the tails.

These thin and thick filaments are aligned in an orderly fashion to make the longitudinal structure known as a sarcomere. Thick and thin filaments overlap along some parts of the sarcomere, but not in others, and this condition of varying degrees of overlapping filaments produces light and dark bands when skeletal muscle is examined with a microscope. The band in which the thick filaments are by themselves is called the **H zone** (Figure 5.1). The area in which the thin filaments overlap the thick filaments is the **A band**; this band also contains the H zone. The area in which there are only thin filaments is called the **I band,** and the ends of the thin filaments are connected to adjacent thin filaments at the **Z lines.** The I band differs from the A band and H zone in that it overlaps the Z line, which demarcates adjacent sarcomeres. Therefore, what is referred to as the I band really encompasses the ends of two adjacent sarcomeres.

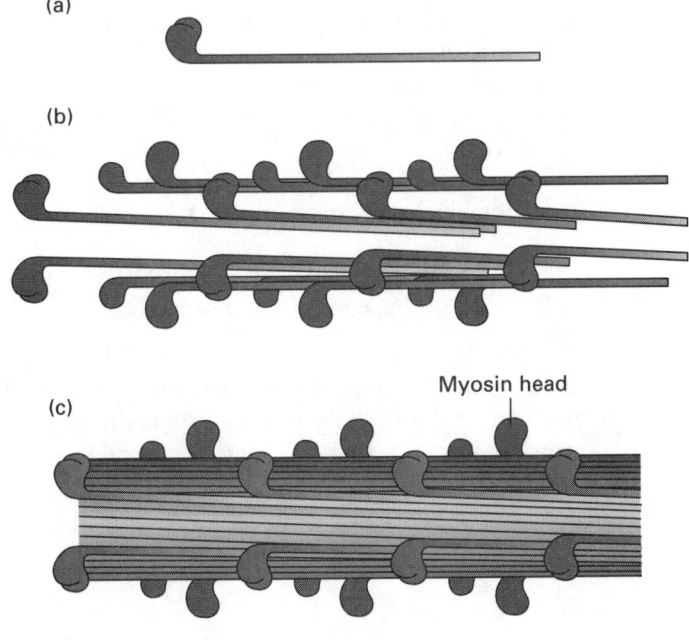

Figure 5.3
The Thick Filament.
An individual myosin molecule can be modeled as a golf club (a). Myosin molecules have a head (club part) and a tail (handle part). The thick filament structure can be modeled as a bundle of golf clubs (b), with the handles all in the same direction. Then, while keeping the handles overlapped, spread the bundle of clubs out so that there are 18 clubs (myosin molecules) at any section at any one time (c).

The Physiology of Muscle Contraction

Muscle movement can be studied at different levels. **Kinesiology** (*kinesis* (G), movement; *logy* (G), study of) deals with what muscle moves a particular part of the body. This is a very important topic in athletics, physical therapy, and human performance. Muscle contraction also can be studied at the level of the whole muscle to understand the relationship of forces, fulcrums, and lever systems to muscle function. Most basically, muscle movement can be studied at the level of the individual sarcomere.

Two additional features of muscle histology not mentioned previously and that cannot be seen readily with light-level microscopy are the **T-tubules** and the **sarcoplasmic reticulum** (Figure 5.4). The T-tubules are extensions of the cell membrane into and through the muscle cell interior. They are numerous, branching, interconnecting tubes of cell membrane that run throughout the muscle cell. The sarcoplasmic reticulum is the specialized, smooth, endoplasmic reticulum of a muscle cell.

Muscle contraction is stimulated by nervous signals that arrive at the muscle cell and initiate the steps leading to movement (Figure 5.5). When an action potential arrives at the **muscle endplate,** the transmitter substance (TS) **acetylcholine** is released from the neuron end. This TS binds to receptors in the endplate and initiates an action potential in the muscle cell membrane. The action potential travels along the membrane and, therefore, down all of the T-tubules as well. (Remember that muscle cells are large cells; therefore, the T-tubules carry the signal for contraction throughout the muscle fiber.) The electrical currents associated with the action potential stimulate the release of calcium ions (Ca^{2+}) from the sarcoplasmic reticulum, where the calcium is stored when the muscle is at rest. (Because calcium is pumped into the sarcoplasmic reticulum continuously, there is very little "free" calcium in the cytoplasm.) Calcium ions are the intracellular "trigger" that "fires" the events of muscle contraction. Just as it is necessary for a gun to be loaded to fire, muscles must be loaded with energy (in the form of ATP) to contract. Also, just as a cocked gun is ready to fire as soon as the trigger is pulled, muscle is ready to contract whenever the stimulus arrives.

The process of muscle contraction can be broken down into three formulas that portray individual steps of the process:

1. Actin + Myosin-ATP → Actin-Myosin-ATP
   Cross-bridge Formation

   The first step in the process is cross-bridge formation. Cross bridges are the heads (golf clubs) of the myosin molecules (of the thick filaments) that have attached to the binding sites on the actin molecules (of the thin filaments). For cross-bridges to form, ATP must be bound to myosin. In muscles at rest, few cross bridges are present.

Figure 5.4
Specialized Organelles in a Muscle Cell.
The sarcomere is the contractile component of muscle. Associated with the sarcomere are the T-tubules and the sarcoplasmic reticulum. The T-tubules carry the action potential throughout the muscle cell, and the sarcoplasmic reticulum is a storage depot for calcium in the cell.

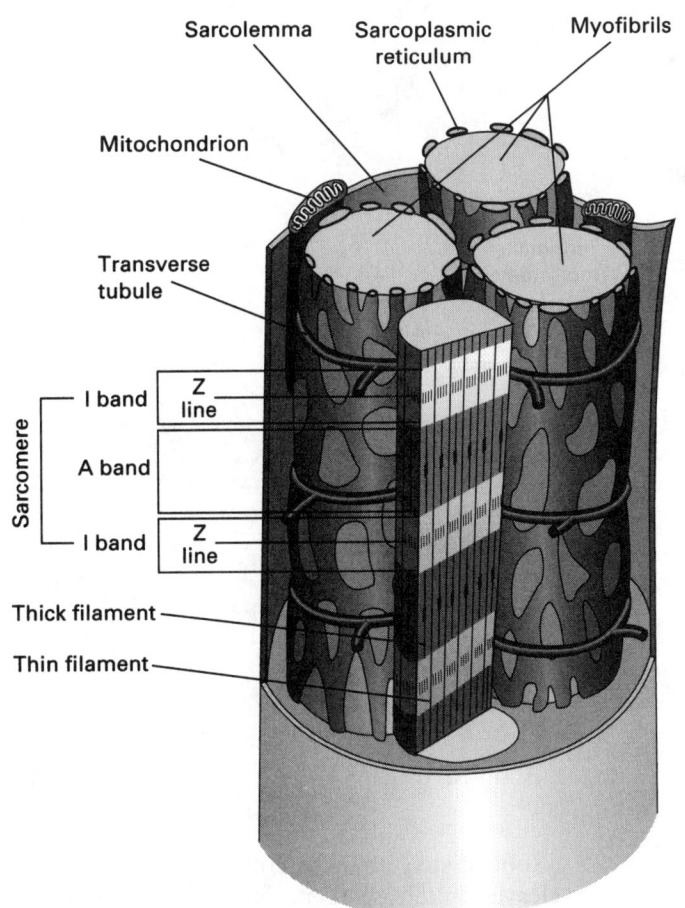

QUESTION: Muscles constantly receive a stream of action potentials, even when the muscles appear to be relaxed. This prevents the muscles from feeling totally limp to the touch. What is this slight state of contraction called? (Hint: When people do not really want to increase the size of their muscles, they might say they just want to increase or improve their muscle _____.)

Observations:

Conclusion:

The Physiology of Muscle Contraction 71

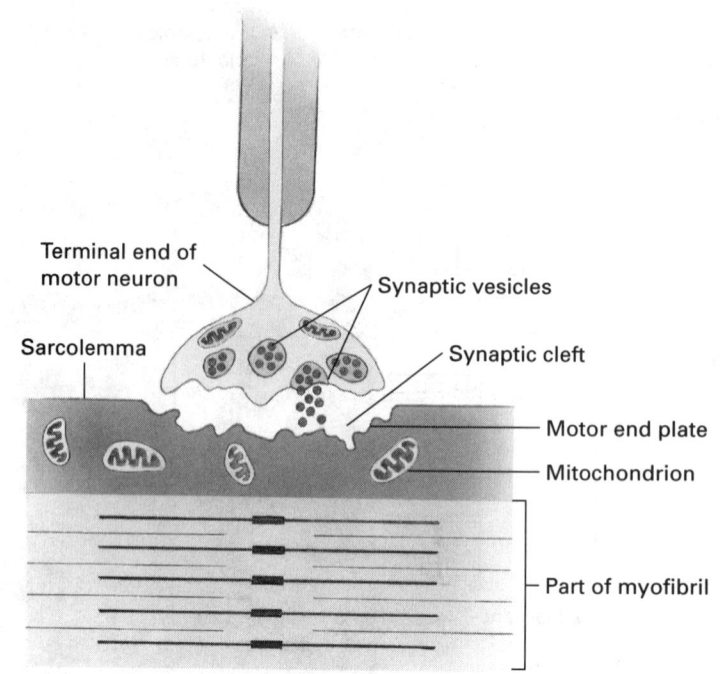

Figure 5.5
The Events Triggering Muscle Contraction.
(1) A muscle is stimulated to contract by an electrical impulse (action potential [AP]) arriving at the muscle on a neuron. (2) The action potential causes the release of a transmitter substance (TS, a chemical signal) from the neuron's end. (3) The TS binds to the receptors on the endplate of the muscle cell. (4) TS binding results in an action potential on the muscle membrane. (5) Muscle contraction occurs.

2.   Actin-Myosin-ATP → Actin-Myosin + ADP + P_i + Energy
Cross-bridge Movement

Immediately after the cross bridge forms in the first step, ATP is automatically cleaved to release adenosine diphosphate (ADP), inorganic phosphate (P_i), and the energy that was present in a high-energy bond. Part of this energy then is used to change the cross-bridge angle so that the thin filament is moved toward the center of the sarcomere. Note that the actin and myosin molecules are still attached to each other; in other words, the cross bridge is still intact. (Most of the energy released is "lost" as heat. In fact, the heat released from muscle activity represents most of the heat that maintains our body temperature at 37°C [98.6°F].)

3.   Actin-Myosin + ATP → Actin + Myosin-ATP
Cross-bridge Dissociation

Muscle contraction actually involves numerous cycles of these formulas. Therefore, the cross bridge must be broken or released so that another can form and the steps repeat. Note that the last half of the third formula is identical to the first half of the first formula. The attachment of a new ATP molecule is necessary to disconnect the cross bridge, but it also prepares the components to repeat the steps.

Muscle Contraction: A Play

Creating models of difficult-to-understand processes can help us to "see" or understand the process. The amount that each sarcomere contracts is very small, but many thousands of sarcomeres contracting in a row results in movement. The events of muscle contraction can be acted out with a row of students, representing parts of a sarcomere, to illustrate the cumulative effect of many individual sarcomeres contracting.

Muscle Contraction

Objectives:
- To develop a working model of muscle contraction.
- To appreciate all the components of muscle contraction.

Materials:
- At least seven volunteers

Procedure:

1. Ask for volunteers, or select students, to make a row of at least seven persons. If the room is too small to line up this many students with their arms extended, either extend the row out a door into the hall or move the entire class into the hall.

2. The odd-numbered students of the row will be the Z lines, with their extended arms being thin filaments. The even-numbered students will be the H zones, with their extended arms being thick filaments.

3. The student at one end is the center or fixed origin of the muscle, and throughout the process, he or she should not move. The other students should move as necessary throughout the activity. This will require everyone's cooperation. You should realize you will need to move your feet and change your position as the "muscle shortens."

4. Each thick filament (the arms of the H-zone students) should grasp the hand of the thin filaments (the arms of the Z-line students). The Z-line students should let their arms be grabbed; they should not do any holding. This step is "cross-bridge formation."

5. Now, the H-zone students should bend or flex their wrists to move their hands closer to their midline while still holding the other students' hands. This step is "cross-bridge movement." This process will be messy, because all the students will need to reposition themselves (except for the end student, who is the center or origin of the muscle). To facilitate this process, the student closest to the immovable end of the muscle should carry out this process first, followed by the other thick filaments after allowing the group to reposition itself.

6. The H-zone students now should release their grasp on the hands of the Z-line students. This step is "cross-bridge dissociation."

7. The process should be repeated until the H-zone students' hands are at the armpits of the Z-line students. (Ask them not to tickle the Z-line students, however. This is a college course, after all.)

QUESTION: The physical limit in this model is the arm length (running into the armpits). Is there some physical limit to actual sarcomeres in muscles? What actually limits the shortening of a sarcomere?

Observations:

Conclusion:

8. If the play acting has worked, note where the movable end of the muscle is. This student should be several feet from where he or she began. In other words, the muscle has shortened. (Be sure to give a round of applause to the students who participated in the play.)

Muscles can only shorten. Because one end is attached to a more stable, immovable structure (i.e., the **origin**), this shortening results in movement of the opposite end (i.e., the **insertion**). The return of a muscle to its original length partly results from its extensibility and elasticity, but more so from there being a different, opposing muscle that pulls the first muscle back to its original length. The muscle that performs a specific movement that is being studied or described is called the **agonist** (*agon* (G), a contest,) or the **prime mover.** The opposing muscle that performs the opposite movement is called the **antagonist** (*ant* (G), against; *agon* (G), a contest). Agonist and antagonist are not absolute terms, however. They describe the muscles that accomplish or reverse a particular action, and are reversed when studying the opposite motion.

The Muscles of the Body

The human body has more than 600 muscles. Without cadavers and extensive models, however, it is impossible to learn all their locations. The muscles that can be seen immediately under the skin are the **superficial muscles.** Once these are identified, they can be removed, after which the **deep muscles** are accessible for identification. Here we concentrate on selected superficial muscles of the body and introduce the terminology associated with the muscle system. The superficial muscles are demonstrated with figures and labeled photographs.

Naming the Muscles

Muscles are named in several ways. Some are named by the action they perform. Therefore, there are **flexors** (*flex* (L), to bend) and **extensors** (*ex* (L), out; *tens* (L), stretched), **levators** (*levator* (L), a lifter), **depressors** (*de* (L), down; *press* (L), to press), **adductors** (*ad* (L), to, toward; *duc* (L), lead), and **abductors** (*ab* (L), away; *duc* (L), lead) (Figure 5.6). Muscles also may be named for their relative size when two or more muscles are located in a specific area. Thus, there are **major** and **minor** pairs, **maximus** and **minimus** pairs, and **longus** and **brevis** pairs, to name the most common terms. Muscles may bear the name of their origin(s) and insertion(s) or be named by the number of origins or insertions. They may be named by the shape of the muscle, as in the **rhomboideus** or **trapezius.** Muscles may be named by where they are located as well. For example, the **abdominus** muscles make up the anterior body wall, whereas the **brachii** muscles are in the arm. In addition, muscles may be named by the bones they overlay. Thus, there is a **frontalis, occipitalis,** and a **temporalis** muscle on the skull, and the **tibialis** and **rectus femoris** are associated with the tibia and femur, respectively. In rare cases, muscles may be referred to by names associated with the occupations in which they are used. The **sartorius** is called the tailor's muscle because it allows us to cross our legs and assume the classic tailor's sitting position (*sartor* (L), tailor). Checking the names of only a few muscles indicates that different muscles are named in different ways; thus, the names presented here are just examples.

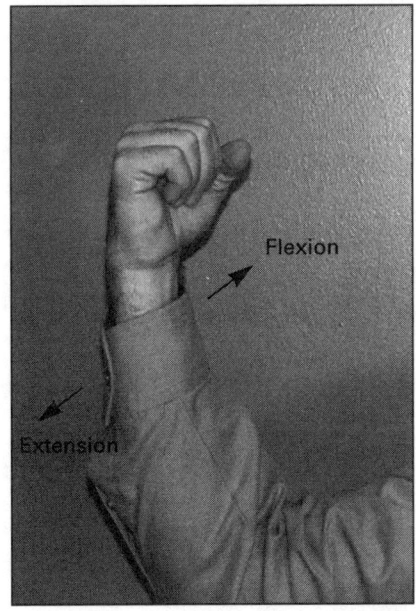 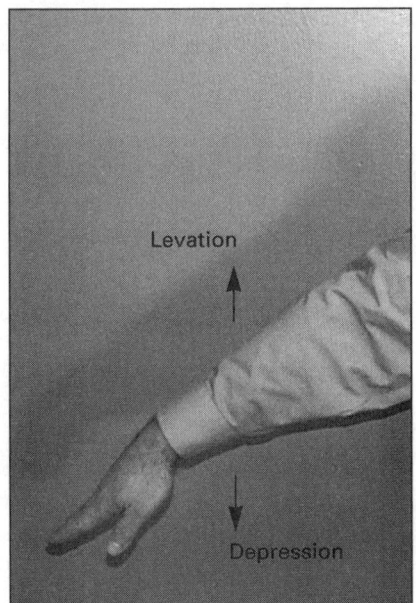 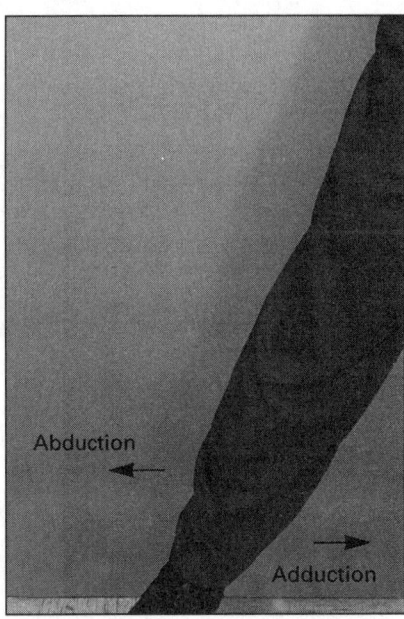

Figure 5.6
Muscle Actions.
Muscles accomplish all actions of the body parts, some of which include (a) extension and flexion, (b) elevation (levators) and pressing down (depressors) and, (c) adduction and abduction.

Learning the Superficial Muscles

Body builders have developed exercises to enlarge most muscles of the body, and muscle magazines or videotapes might help you to learn the superficial muscles. Figures 5.7 through 5.14 show the muscles to learn. They also depict the same areas of the body, with the muscles labeled to help you appreciate where they are in relation to the surface of your body.

Journal Note Observing Your Muscles. Some of us are "built," and some of us are not. Most people, however, can see that as they move their fingers, their forearm moves as well. Roll up your sleeves to expose your forearm. With your hand (palm down) and arm flat on a table, lift each of your fingers one at a time, and note the movement of the different muscles in the arm. Try to name these muscles by which muscle moves which finger. Also, try to identify these muscles with the help of your text or a more specialized source. For example, note the tension in the tendons/muscles surrounding the ankles and the upper leg as, while standing, you rock back and forth, thus shifting your center of gravity. Can you make these same observations for other parts of the body?

Figure 5.7
Muscles of the Face.
(a) Schematic representation.
(b) Photographic representation.

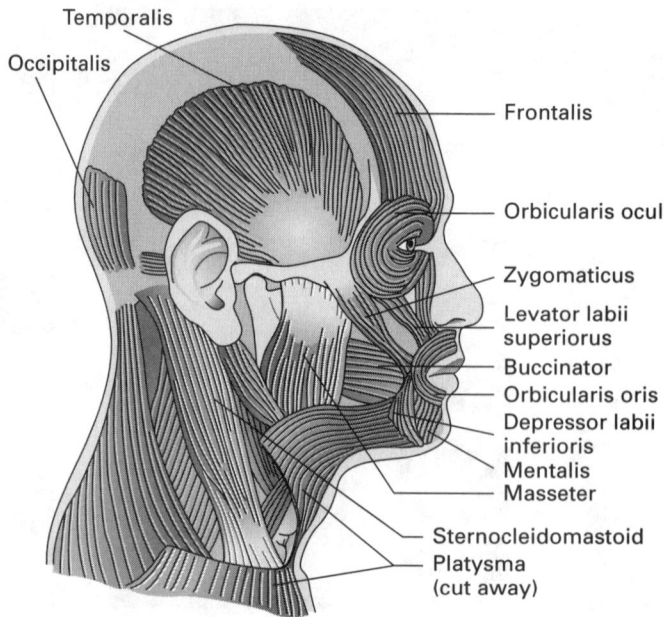

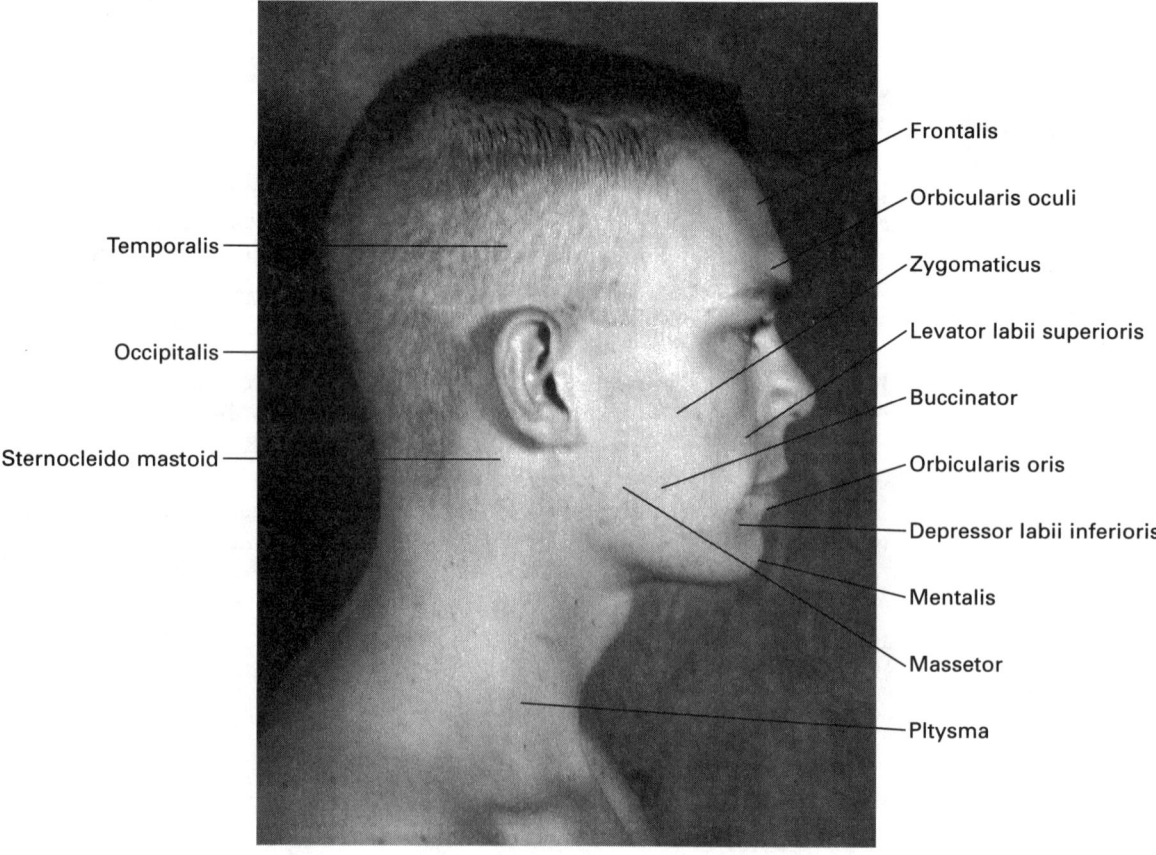

Figure 5.8
Muscles of the Chest.
(a) Schematic representation.
(b) Photographic representation

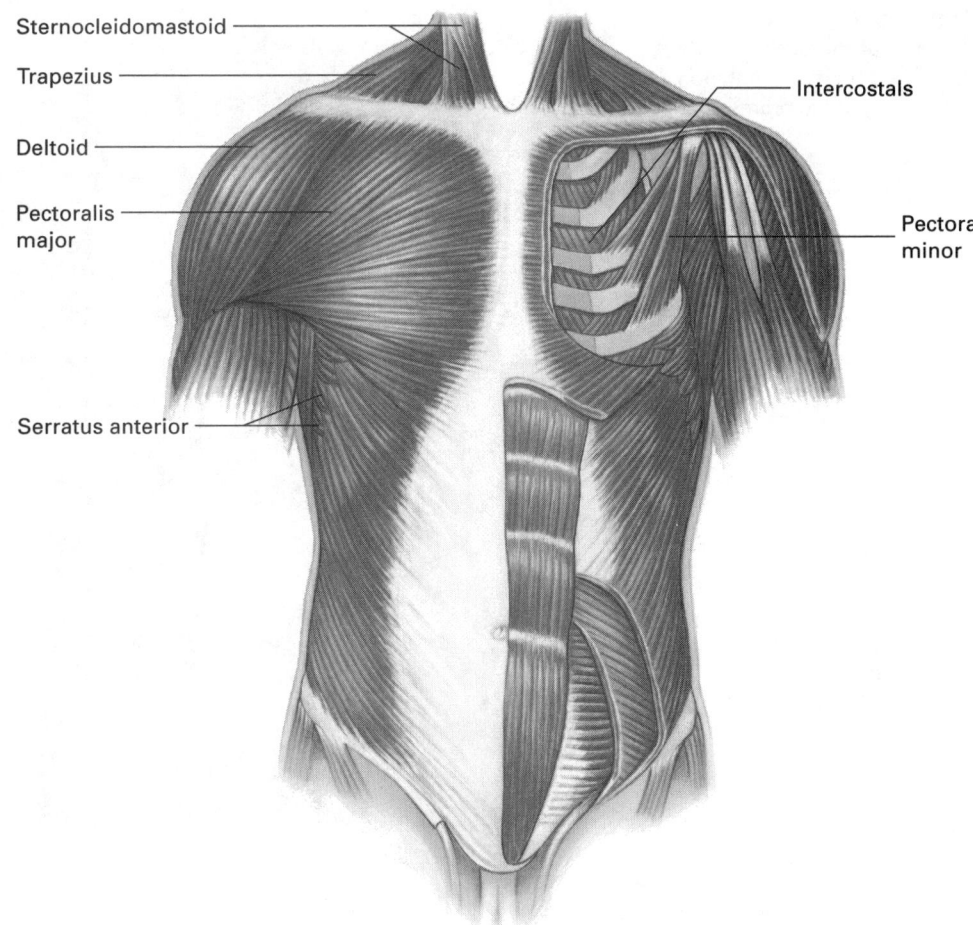

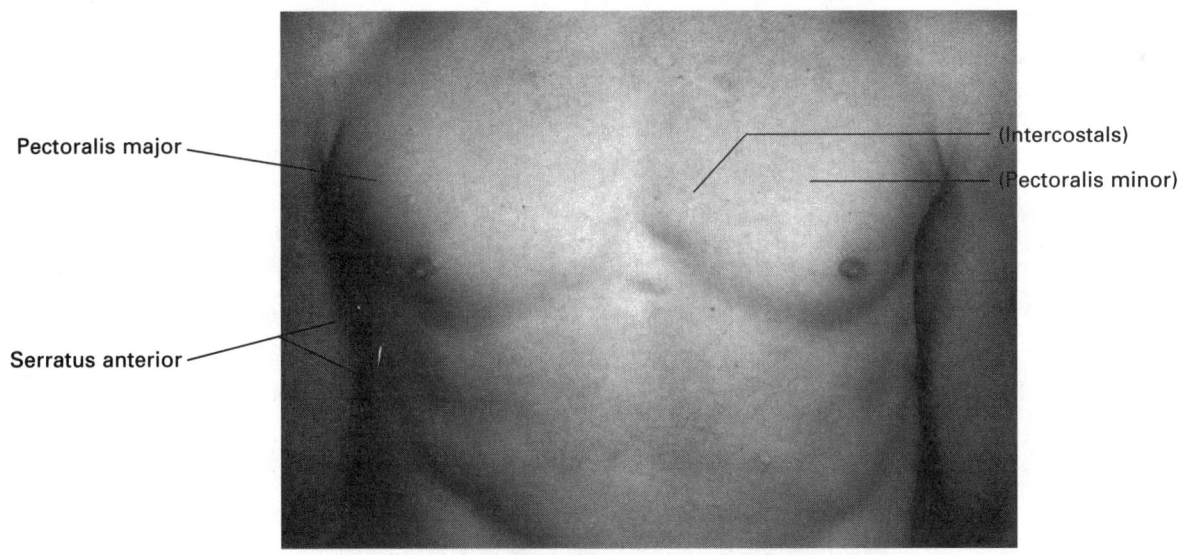

Learning the Superficial Muscles 77

Figure 5.9
Muscles of the Back.
(a) Schematic representation.
(b) Photographic representation.

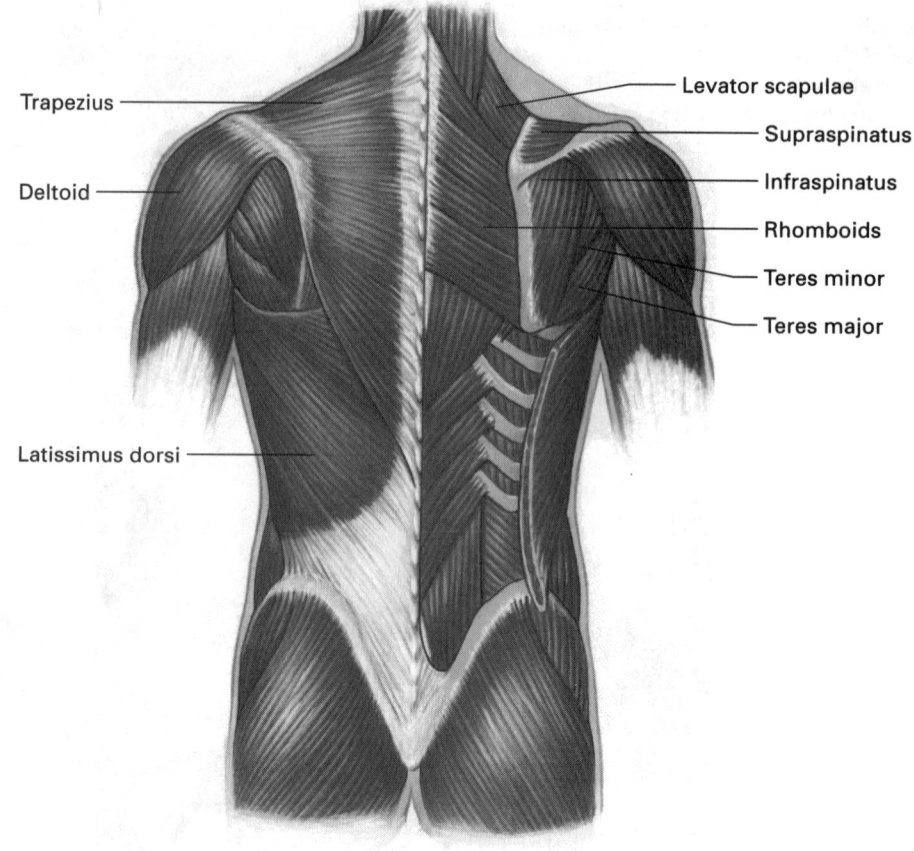

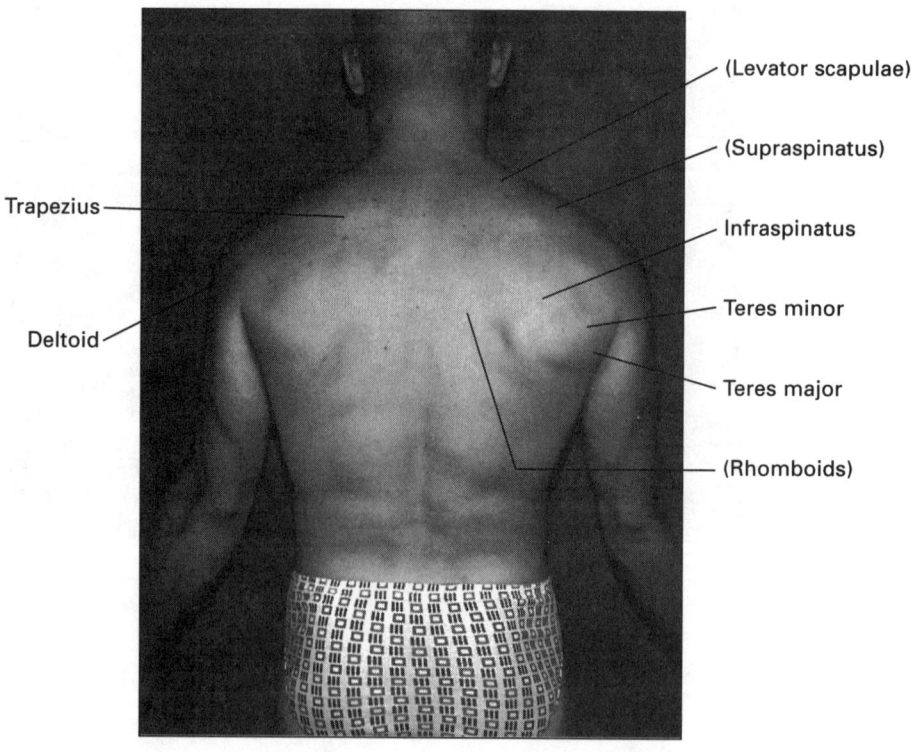

78   SECTION 5   The Muscular System

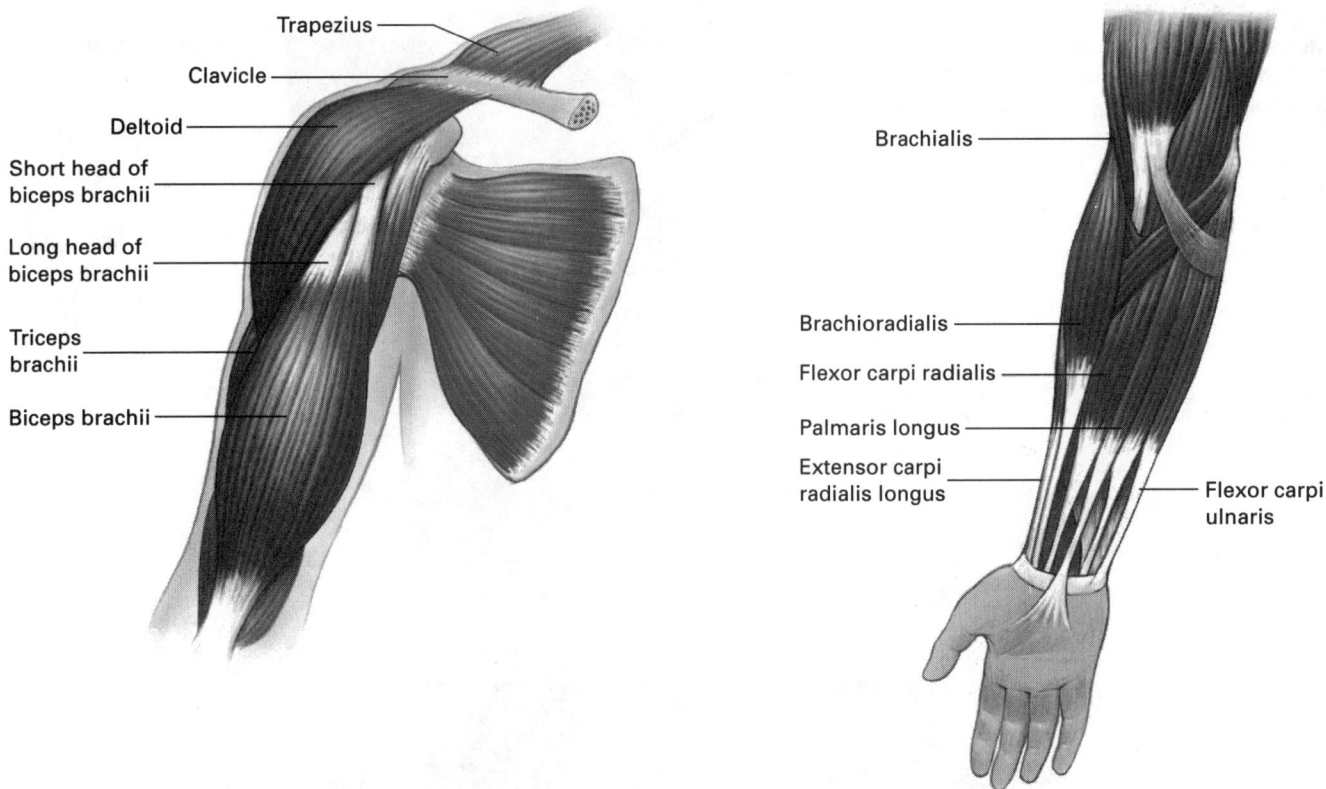
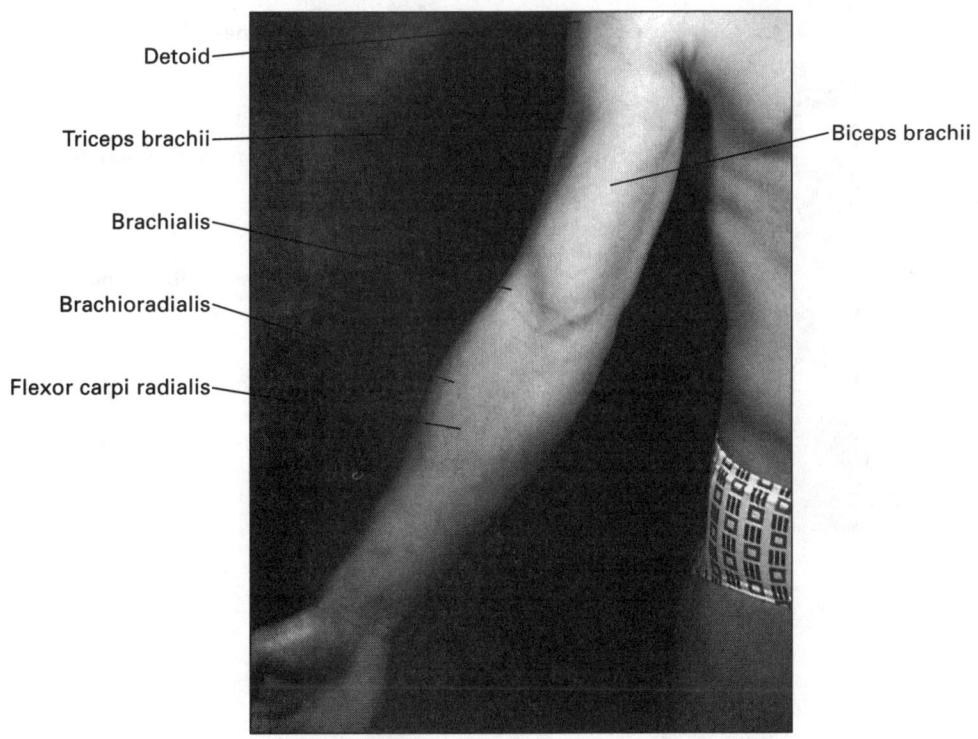

Figure 5.10
Muscles of the Anterior Arm.
(a) Schematic representation. Note there are numerous other muscles in the forearm and hand that are deeper than those described and much smaller then those labelled.
(b) Photographic representation.

Learning the Superficial Muscles

Figure 5.11
Muscles of the Abdomen.
(a) Schematic representation.
(b) Photographic representation.

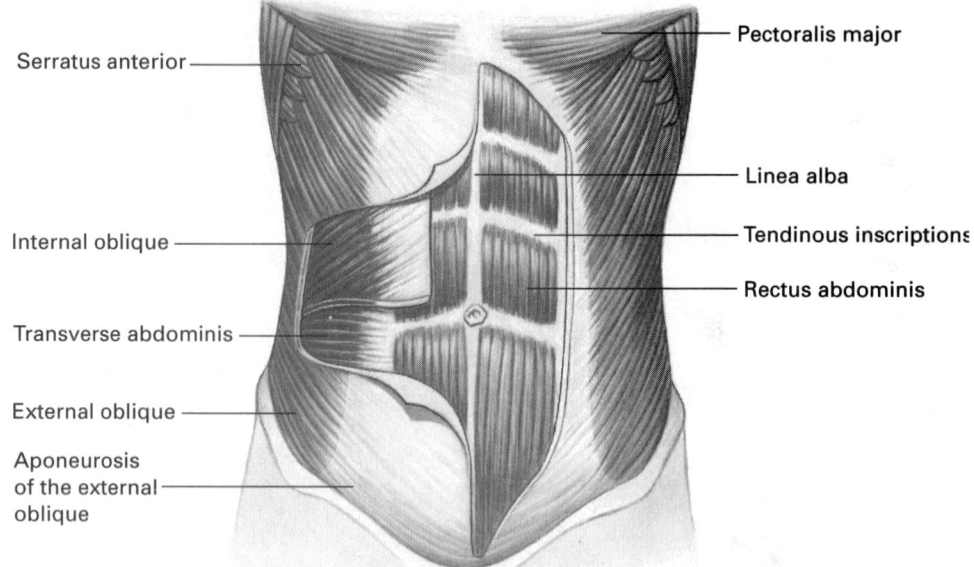

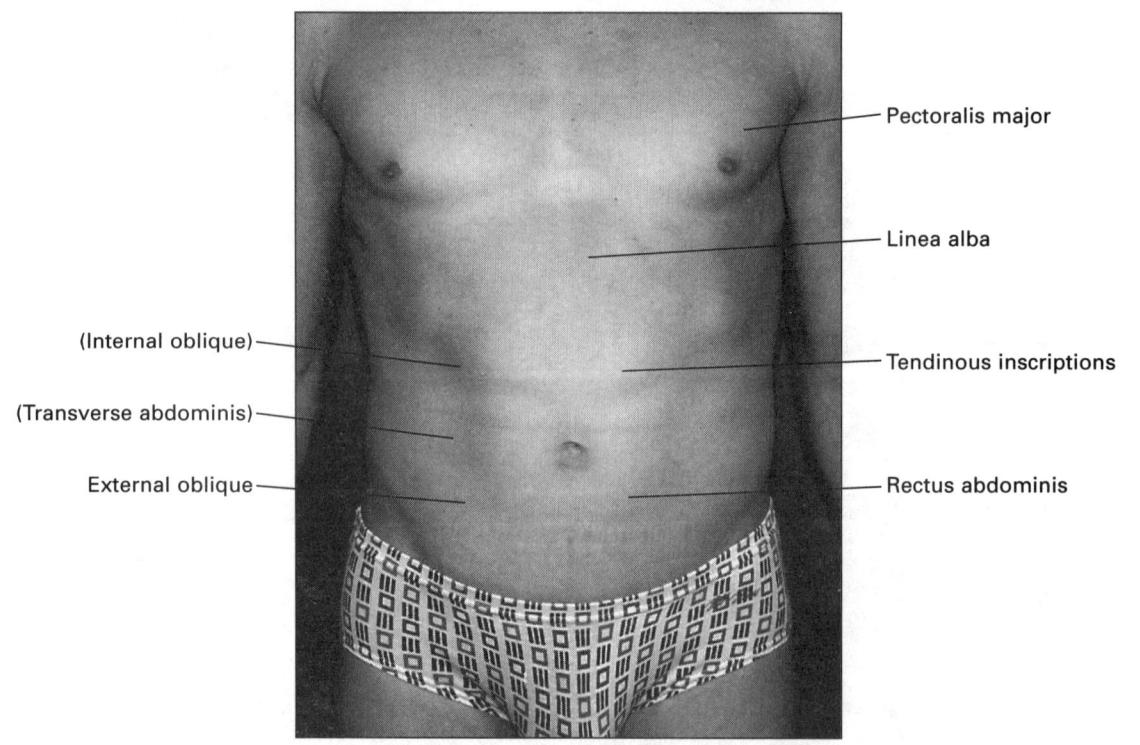

Figure 5.12
Muscles of the Posterior Hip and Thigh.
(a) Schematic representation.
(b) Photographic representation.

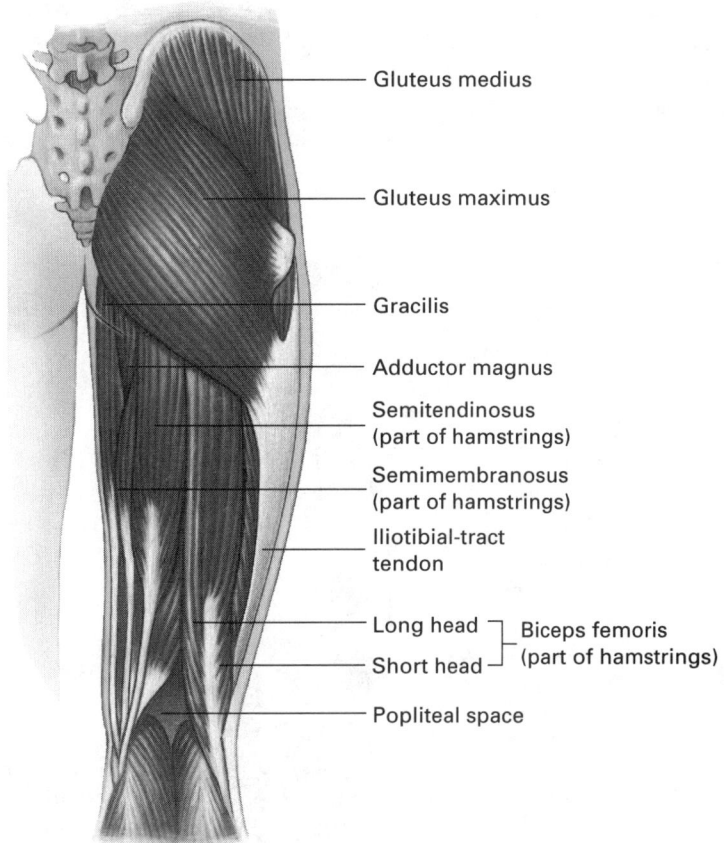
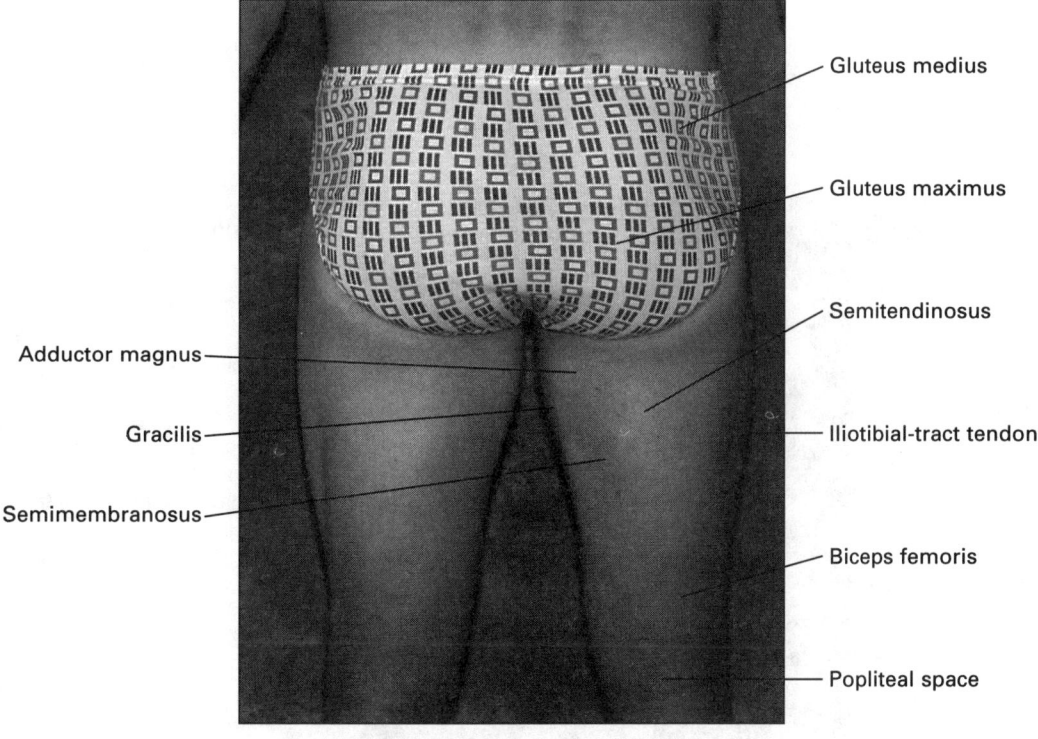

Figure 5.13
Muscles of the Anterior Thigh.
(a) Schematic representation.
(b) Photographic representation.

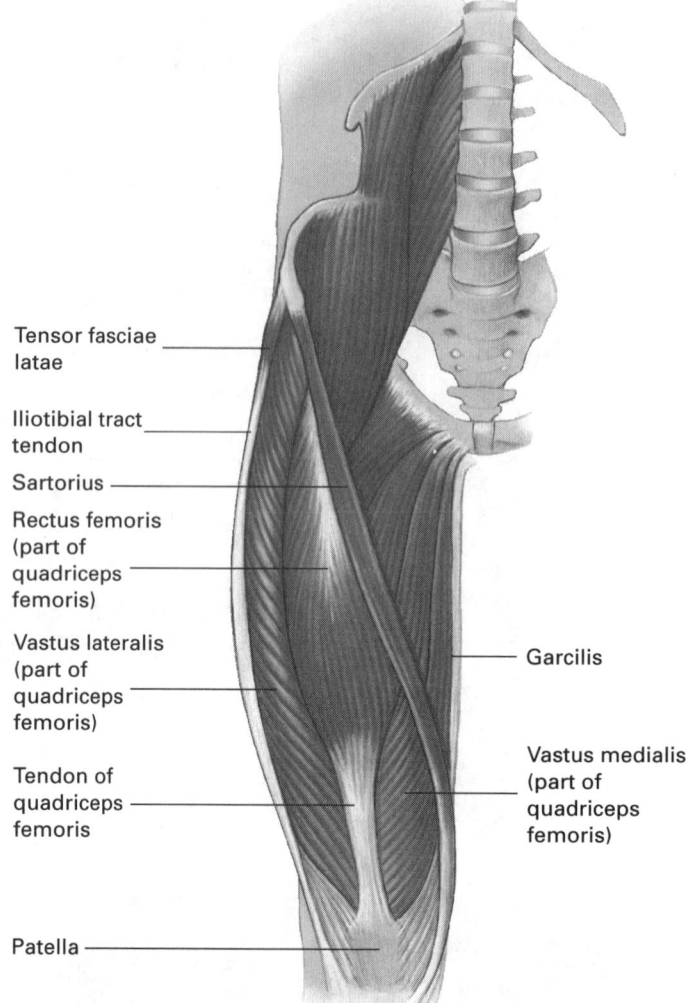

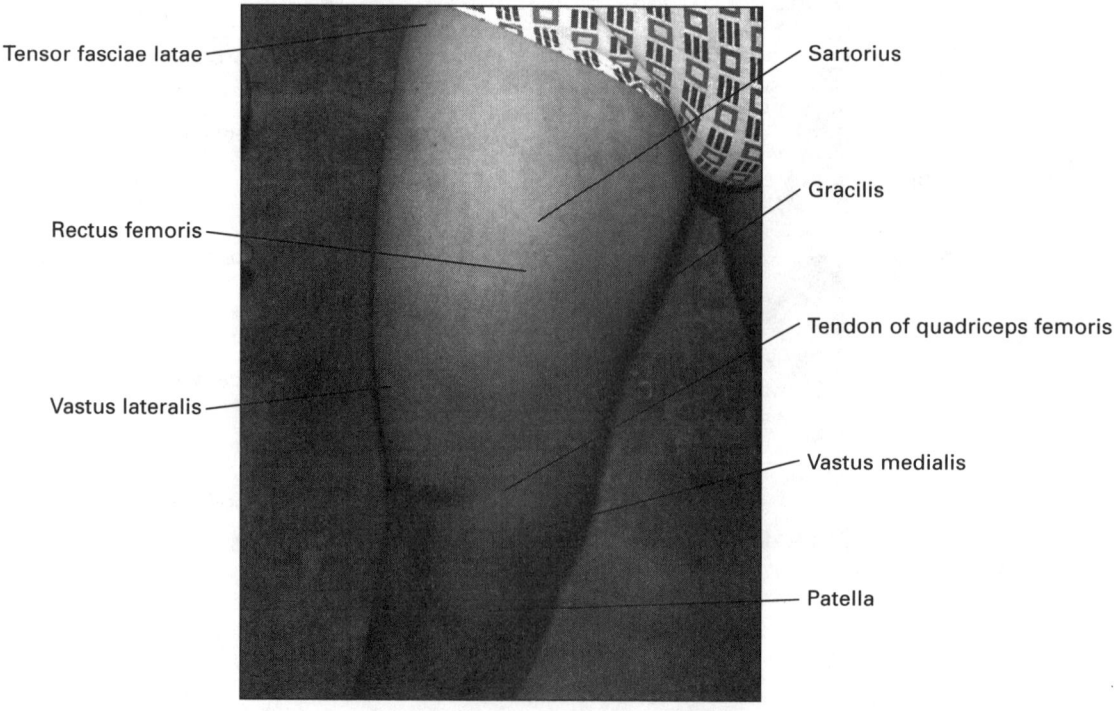

82 SECTION 5 The Muscular System

Figure 5.14
Muscles of the Lateral Lower Leg.
(a) Schematic representation. Note there are numerous other muscles in the lower leg and foot that are deeper than those described and much smaller than those labelled.
(b) Photographic representation.

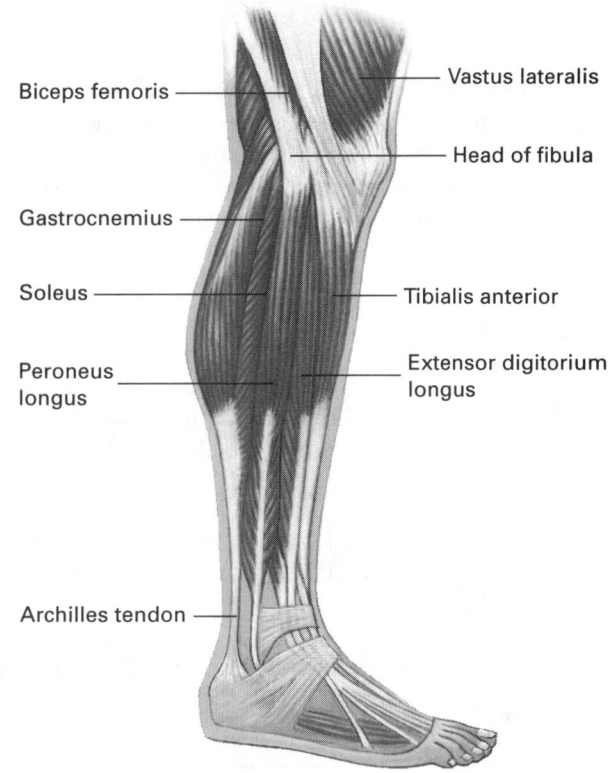

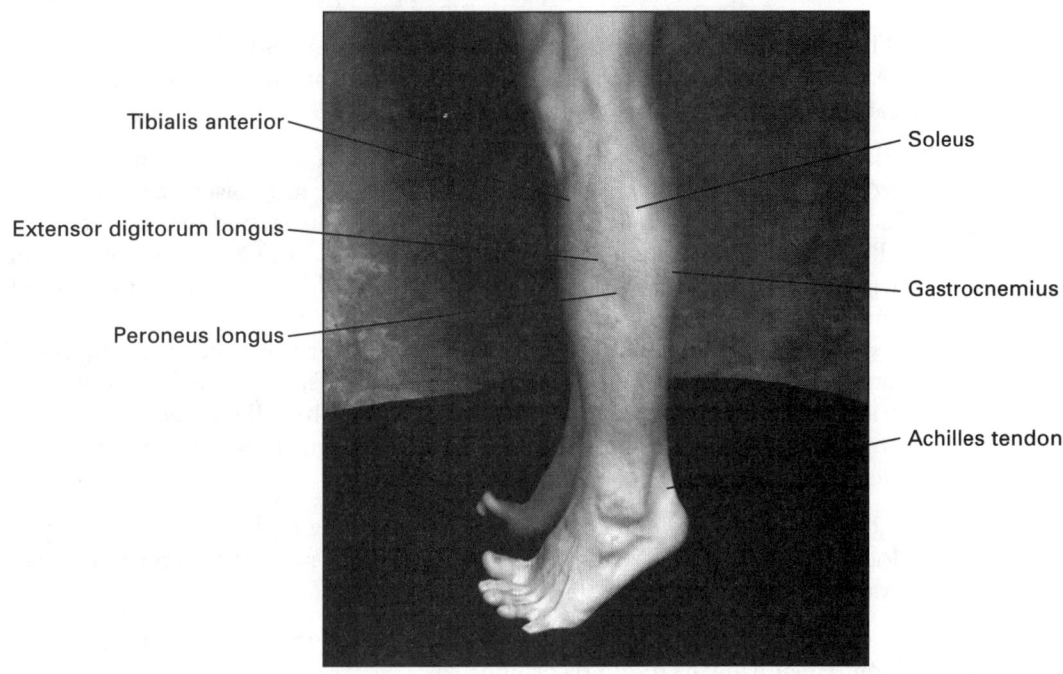

Learning the Superficial Muscles

Electromyography

Many parts of the body have electrical activity associated with them. The retina of the eye, for example, is electrically active; therefore, its activity can be recorded. In fact, recordings from the body surface can be made at any part that is active. You probably are familiar with the electrocardiogram (or ECG) that is recorded from the heart. Because muscles are characterized by electrical activity, they can be recorded from the surface as well.

Electrogram

Objectives:

- To record the electrical activity of a selected muscle or muscle group.
- To appreciate that muscles are electrically active as they contract.

Materials:

- Oscilloscope or physiological recorder with high-gain coupler.
- Electromyographic (EMG) plate electrodes with attachment devices.
- Alcohol pads.
- Electrode paste.
- Small barbell set or small, compact objects of different weights.

Procedure:

Caution: Though not essential, it is good lab procedure to wash your hands before starting this exercise.

---

1. The upper or lower arm represents a good area to study, because it generally is readily accessible. Clean the area to be tested with alcohol to ensure better electrical contact.

2. Place the electrode at the two ends of the muscle selected. If you are recording from the biceps on the anterior surface of the upper arm, place one electrode closer to the shoulder, near the proximal end of the muscle, and the other toward the elbow, near the distal end of the muscle (Figure 5.15). Contraction of the biceps will flex the arm. A third electrode or ground should be placed on the back side of the arm.

   Electrodes may be disposable or reusable. If reusable, they typically contain a small well that must be filled with electrode paste. This paste makes the actual electrical connection with the skin. There also may be adhesive washers or elastic straps available for attachment to the subject. Your instructor will describe the particular set-up available.

3. Connect the electrodes with the appropriate cabling to the oscilloscope or physiological recorder. Observe the signal, and if possible, record when the muscle is contracted to flex the arm.

4. If recording from the biceps, ask the subject to perform the opposite action: extension. What happens to the recording?

5. Ask the subject to lift objects of different weights, and observe any changes in the size and pattern of the recording.

6. Switch the electrodes, including the ground, to the opposite side of the arm. Perform the same procedure as described. How does this recording differ from the previous one?

Figure 5-15
Placement of Electrodes at Two Ends of a Selected Muscle.
If you are recording from the biceps on the anterior surface of the upper arm, place one electrode closer to the shoulder, near the proximal end of the muscle, and the other toward the elbow, near the distal end of the muscle. The ground should be placed on the opposite side of the arm.

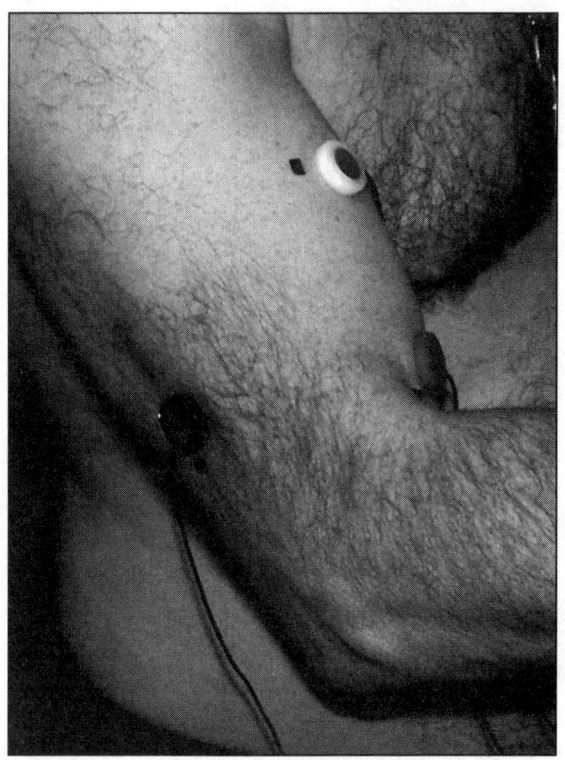

QUESTION: This recording is known as the electromyogram or EMG and it can be recorded from most parts of the body to study the activity of specific muscles or muscle groups. For example, during sleep studies, an EMG of the external muscles of the eye can be recorded to obtain an understanding of eye movement during sleep. What do the eyes do during sleep? (Hint: Watch a baby during sleep.)

Observation:

Conclusion:

Journal Note After working out, muscles tend to be larger and more prominent, and the superficial veins protrude. Try it. Why are they larger and more red after exercising?

Electromyography

Section 6

The Endocrine System

Note to the Student

The endocrine system consists of the various ductless glands throughout the body that produce hormones (Figure 6.1). It is the "other" regulatory system, in addition to the nervous system. The nervous system carries out those responses that require immediate action, whereas the endocrine system usually is responsible for longer-term regulatory schemes. Beacause these endocrine glands do not have ducts, they secrete their products (i.e., the hormones) directly into the bloodstream.

A definition of a **hormone** may help at this point. A hormone is a chemical that is produced by specialized cells in a specific location of the body, that is released into the bloodstream in very low concentrations, that circulates throughout the body, and that only affects certain "target" organs or cells. (What are the most important parts of this definition?) Hormones are not chemicals that are produced by any cell; they are only produced by specialized cells in a certain locations. They are released into the bloodstream and, therefore, not into a body cavity or onto a surface. Hormones are very specific and effective, so even though they are distributed generally throughout the body, a small quantity is adequate to produce the desired effect. Also, though the hormone is in contact with all cells, only those that are targets are responsive.

QUESTION: What is there about a cell or tissue that would make it responsive to a hormone?

Observations:

Conclusion:

The definition of a hormone just given is a very classical one. The concepts and principles of hormone function have been modified since the early days of endocrinological research, however. The range of chemicals now considered to be hormones has been broadened.

Figure 6.1
The Endocrine Glands of the Body.
The major endocrine glands of the body are shown at their respective locations. Other endocrine tissues throughout the body are not shown. (These may be mentioned in discussions of the particular system whose activities they affect.)

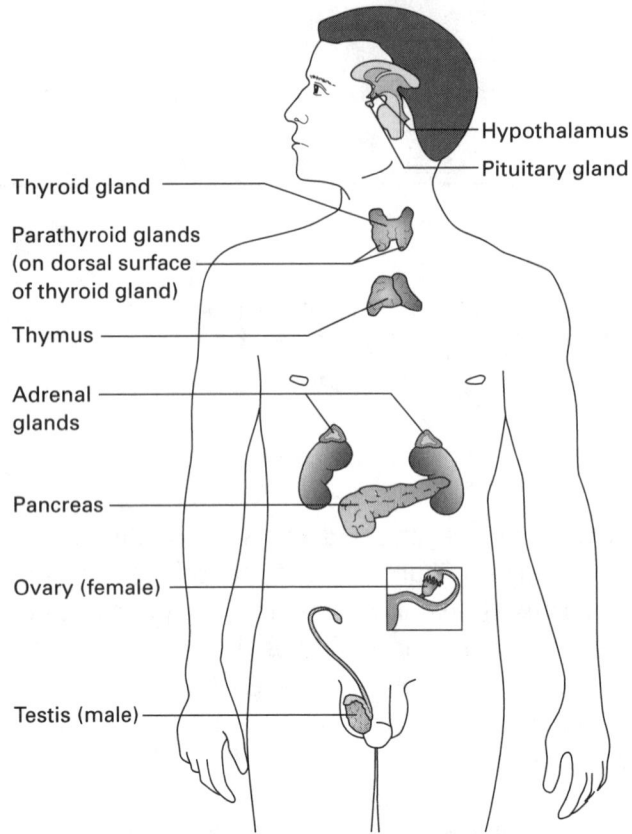

Journal Note Hormones are messengers that help to coordinate some very important processes in the body. If people produce too much or too little of a particular hormone, they may experience some disturbing signs and symptoms. What are some conditions for which people receive replacement hormones or other treatments or medication to adjust their hormonal function? (Hint: Consider diabetes mellitus and insulin, menopause and estrogen, development and growth hormone, and thyroid conditions.) Even if a condition has a hormonal basis, we sometimes cannot—or choose not to—treat the problem hormonally. What are some of these conditions and the explanation for this choice? (Hint: Consider early stage insulin-dependent diabetes, non-insulin-dependent diabetes mellitus, and hypoglycemia.)

Types of Hormones

Two types of hormones can be distinguished based on their solubility. The protein and protein-like/amino acid hormones are **water soluble**; the lipid/steroid hormones are **lipid soluble**. This basic feature of hormone structure affects how these hormones carry out their actions. The protein hormones cannot get through the cell membrane and, therefore, must have a means to carry out their action inside the cell (Figure 6.2). The water-soluble hormones have their effect through a receptor in the membrane. The receptor is linked to an enzyme on the inner side of the membrane. When the hormone is present and bound to the receptor, it activates the enzyme that stimulates production, inside the cell, of a **second messenger** to carry out its effect. In this sense, the hormone is the **first messenger.**

Lipid-soluble hormones can readily enter all cells of the body. In those cells that are responsive to the hormone, they combine with a receptor in the cytoplasm to form a hormone-receptor complex, which enters the nucleus and turns on expression of the specific genetic information of the cell.

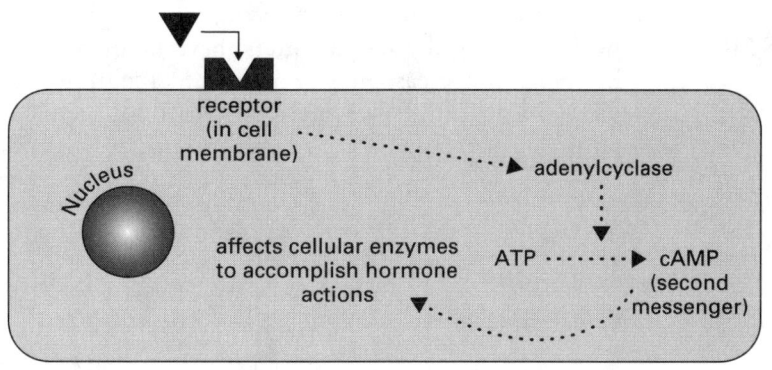

Figure 6.2
The Hormone Receptor.
The cell membrane of a target cell contains receptors for water-soluble hormones. The hormone actually activates an enzyme that is part of the receptor to convert a substrate into a product that signals the hormone's effect inside the cell. (*ATP,* adenosine triphosphate; *cAMP,* cyclic adenosine monophosphate.)

Control of Hormone Release

Hormone release and many other metabolic processes are regulated through **negative-feedback mechanisms.** A negative-feedback system is one in which the level of the hormone being released feeds back to tell the producing gland or its controller to decrease production or release of the hormone in question. This type of regulatory scheme is very important in homeostasis, because it helps to hold parameters within intended ranges. If the level is too high, this information feeds back to the controller to lower it. If the level is too low, the system responds by raising the level to maintain the needed concentration.

The Pituitary Gland

The **pituitary gland** has more accepted names for its various parts than most other structures in the body. It also is known as the **hypophysis.** Located at the base of the brain and attached to the hypothalamus by a stalk called the **infundibulum,** it is protected by the **sella turcica** of the sphenoid bone, which wraps up around it. It is divided into an **anterior lobe,** which also is called the **adenohypophysis,** and a posterior lobe. The anterior lobe produces and releases the hormones listed in Table 6.1.

Table 6.1 Hormones Released from the Anterior Pituitary Gland

Name	Abbreviations
Growth hormone, somatotropin	GH, STH
Thyroid-stimulating hormone	TSH
Adrenocorticotropic hormone	ACTH
Melanocyte-stimulating hormone	MSH
Prolactin	PRL
Follicle-stimulating hormone	FSH
Luteinizing hormone	LH

QUESTION: Athletes, especially weight lifters, have long abused anabolic steroids (i.e., male sex hormone) to increase muscle mass and strength for competition. Of the hormones in Table 6.1, which (if any) might be abused for a similar reason?

Observations:

Conclusion:

The pituitary gland also has a **posterior lobe** that serves only as a point of release for hormones rather than as a site for their production; it also is called the **neurohypophysis.** The posterior lobe has a different embryonic origin than the anterior lobe as well. Whereas the posterior lobe is a downward (i.e., inferior) extension of the "floor" of the hypothalamus, the anterior lobe grows from Rathke's pouch, which grows up (i.e., superiorly) from the "roof" of the mouth and, eventually, completely separates from it. The hormones released from the posterior lobe are listed in Table 6.2.

Journal Note Most parents are concerned their children receive proper nutrition so they can grow and develop to their full potential. One of the tools that is used to measure the normalcy of growth is the growth chart, in which the height and age of a child is compared with those in the chart to determine into what percentile the particular individual fits. Over the last 10 years, there has been interest in using recombinant DNA growth hormone to treat children who are not keeping up with their same-age cohort on the growth chart. What do you think about the ethics of manipulating a child's growth through hormone injections? How do we really know a child is not growing as fast as he or she should be?

The Thyroid Gland

The **thyroid gland** has a distinct internal structure. It contains spheres called **follicles,** which are lined with cuboidal epithelium, that secrete the hormones **thyroxin** (T_4) and **triiodothyronine** (T_3). (The subscript indicates the number of iodine atoms in this form of the hormone.) The hormones are either released directly into the bloodstream or stored temporarily in the central cavity of the follicle as **colloid,** which is a clear, viscous fluid. Because the follicles are spherical in shape, there are spaces

Table 6.2 Hormones Released from the Posterior Pituitary Gland

Name	Abbreviation
Oxytocin	OT
Antidiuretic hormone, vasopressin	ADH

between the follicles when they are packed together in the gland. These spaces are filled with blood vessels and **parafollicular cells,** which produce another hormone, **calcitonin,** that helps the body to regulate the calcium and phosphate levels in the bloodstream.

Thyroid Gland

Objectives:

- To learn the overall structure of the thyroid gland.
- To appreciate that more than one hormone can be produced by different areas or cells in the same gland.

Materials:

- Microscope.
- Thyroid slide, section.

Procedure:

1. Obtain a slide of the thyroid gland.

2. Note how the gland has the appearance of various-sized, circular structures that are packed together (Figure 6.3). These are the follicles. They are spherical structures lined with cuboidal and columnar epithelial cells that produce T_3 and T_4 and store it as colloid in the antrum.

3. The cells located outside the follicles are the parafollicular cells. These make calcitonin, which serves in calcium regulation.

QUESTION: Calcitonin actually increases the blood level of calcium. Considering the body's principal stores of calcium are in bone, what do you think calcitonin does to the calcium stored there?

Observations:

Conclusion:

Figure 6.3
The Thyroid Gland.
This representative section of thyroid shows the follicles of the gland that produce thyroid hormone.

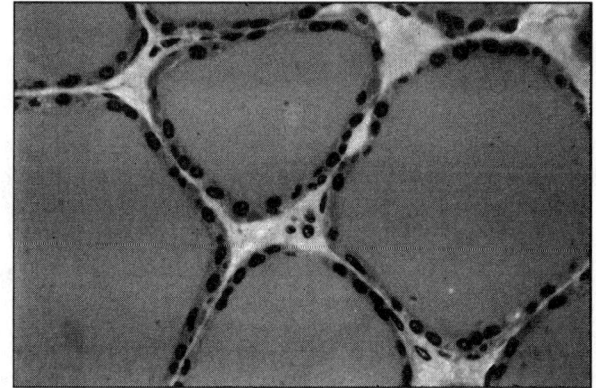

QUESTION: The thyroid gland is the only organ in the body that uses iodine, which is the radioactive substance most likely to be released during accidents at nuclear power plants (i.e., "hot" iodine). What health risks are likely to be caused?

Observations:

Conclusion:

The Parathyroid Glands

The **parathyroid glands** are four to five small, pea-shaped masses that are embedded in the posterior surface of the thyroid gland. These glands are very simple in structure, and they release only one hormone, **parathyroid hormone (PTH),** which also influences the blood calcium and phosphate levels.

QUESTION: One research technique that was used by early endocrinologists is called ablation (*ablat* (L), remove). In this technique, the researcher would remove a suspected endocrine gland and observe changes in the animal subject. Can you hypothesize about the problems with these results if the parathyroid glands were accidentally removed with the thyroid?

Observations:

Conclusion:

The Adrenal Glands

The **adrenal glands** (*ad* (L), toward; *rena* (L), a kidney) sit atop the superior ends of each kidney. The glands contain an inner **medulla** and an outer **cortex,** with each having a different function (Figure 6.4). The medulla actually is a sympathetic ganglion, so it is similar in its secretions to those produced by the sympathetic autonomic nervous system. The medulla releases both **epinephrine** and **norepinephrine,** both of which help to prepare the body to deal with stressful situations. The adrenal cortex is a steroid-producing gland, and Table 6.3 lists the hormones it produces.

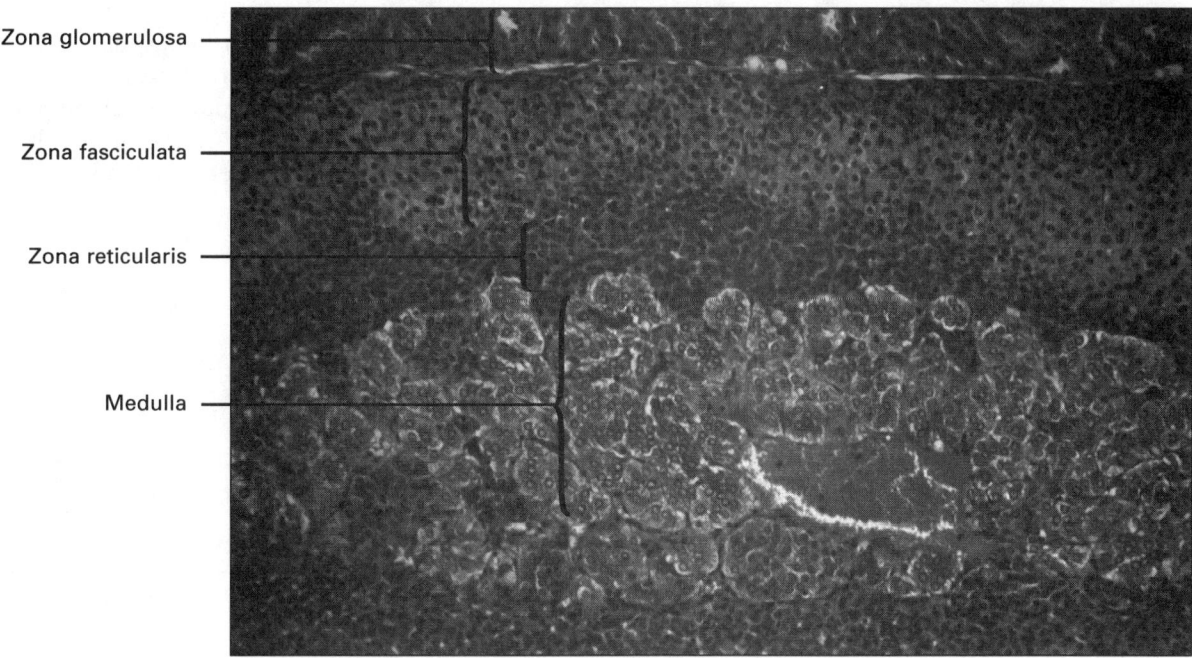

Figure 6.4
The Adrenal Gland.
This representative section of adrenal gland shows the two principal regions of the gland: the cortex, and the medulla. The adrenal gland originates from two different embryonic tissues, thus producing the dual structure of the gland.

Typically, we associate the sympathetic autonomic nervous system and adrenal medulla with helping the body respond to stressful stimuli, but steroid hormones released by the adrenal cortex also are involved with the body's ability to deal with long-term stress. To improve our chances of survival in dangerous or threatening situations, the body has what is known as the "fight or flight" response. This automatic set of reflexes allows a person to deal with a stressful lifestyle.

Table 6.3 Hormones Released from the Adrenal Cortex

Name	Role
Mineralocorticoids	
Aldosterone	Sodium and potassium regulation
	Blood volume regulation
	Blood pressure regulation
Glucocorticoids	
Cortisol	Glucose and glycogen synthesis
Corticosterone	Increase blood sugar
Cortisone	Reduction of inflammation
Sex hormones	
Androgens	Male sex hormones
Estrogens	Female sex hormones

The Adrenal Glands

QUESTION: Everyday demands often tax our abilities to deal with life's situations. In the terminology of experts in this field, stress that is damaging to our systems is called distress. However, certain demands of life, though disruptive and painful, help us to prepare for real crises, and this type of stress is called eustress (*eu* (G), good). Eustress is like exercise. If exercise helps our systems to become more efficient and capable, what does eustress do to our "fight or flight" response?

Observations:

Conclusion:

Adrenal Glands

Objectives:
- To learn the structure of the adrenal gland.
- To understand the dual nature of the adrenal gland in its origin and function.

Materials:
- Microscope.
- Adrenal slide, section.

Procedure:

1. Obtain a slide of the adrenal gland.

2. There should be two different regions on the slide (Figure 6.4). The surface area is the adrenal cortex, and the inner area is the adrenal medulla.

3. The cortex is not a homogeneous arrangement of cells. Three layers of varying arrangements are associated with producing the different hormones the adrenal cortex manufactures.

4. The outermost layer of the cortex is called the **zona glomerulosa** because of the clustering of the spherical groups of cells. The second layer is made of linear cell cords and is called the **zona fasiculata.** The innermost layer, which is adjacent to the medulla, is called the **zona reticularis,** because the cells here are arranged as in a mesh or a net.

5. The medulla contains clusters of **chromaffin cells,** which produce hormones in response to stimulation by the sympathetic autonomic nervous system.

QUESTIONS: The autonomic nervous system is arranged as a preganglionic neuron and a postganglionic neuron, which attaches to some effector. Innervation of the adrenal gland is different from that of other parts of the body. The preganglionic fiber connects directly to the adrenal gland. What does this unusual arrangement tell you about the nature of the adrenal medulla?

Observations:

Conclusion:

The Pancreas

The **pancreas** is somewhat unique, because it is an endocrine as well as an exocrine gland. Exocrine glands secrete into a duct or cavity rather than into the bloodstream. The two functioning aspects of the pancreas are separate. The endocrine component is the **islets of Langerhans;** the exocrine component is the **pancreatic acini.** Islets in the pancreas appear as endocrine islands in the larger body of the exocrine gland and they contain beta cells, which secrete **insulin,** and alpha cells, which produce **glucagon.** Both of these hormones are involved in the regulation of blood glucose concentration.

Determination of Blood Glucose Concentration

The regulation of blood sugar by competing hormones probably is one of the most studied and physiologically important blood parameters. Adequate and reliable blood sugar levels must be maintained for proper brain function. Insulin is released after a meal when the body is digesting foodstuffs and absorbing large amounts of sugar. Because there is an upper limit on the amount of glucose that kidneys can process, the opposing hormones are designed to prevent the blood glucose level from climbing above this level. Insulin helps to prevent the rise in blood sugar by stimulating liver and skeletal muscle cells to absorb glucose and convert it into glycogen for later use.

Glucagon is the opposing hormone. When the blood sugar level begins to fall, glucagon is released and stimulates liver cells to convert glycogen into glucose and release it into the bloodstream. These mechanisms prevent the blood sugar level from getting too high or too low for proper brain function, and this competitive regulatory scheme between insulin and glucagon is an often-cited example of homeostasis in action.

Diabetes mellitus occurs when the body cannot adequately regulate the blood sugar level. There are different forms of diabetes based on the patient's age at onset and the sensitivity of the condition to treatment with insulin. To determine whether a person has diabetes, a fasting blood sugar determination is made and a glucose tolerance test performed. Clinical presentation and an elevated glucose level, however, usually are sufficient to diagnose the insulin-dependent form.

Recent advances in the treatment of diabetes have produced many new ways to easily determine blood glucose levels. Because high levels of glucose "spill" into the urine, early methods involved chemically treated strips, which were inserted into a sample of urine. Because this provided only an indirect measurement, however, regulation of blood sugar by insulin injection was poor. Today, minispectrophotometers are available at very affordable prices that allow direct and frequent measurement of blood sugar levels.

Determination of Blood Glucose Concentration

Objective:

- To understand the importance of monitoring glucose in the treatment of diabetes.
- To appreciate the ease of operation and ready availability of blood glucose monitoring devices.
- To compare recent diet and blood glucose level to illustrate homeostatic mechanisms at work.

Materials:

- Blood glucose monitoring device (available from scientific and hospital supply companies and local pharmacies).
- Treated strips for the glucose monitoring device.
- Alcohol pads.
- Lancets (disposable), or automatic lancing devices with disposable lancets.
- Biohazard disposal containers.
- Fresh, 10% bleach solution.

Procedure:

1. Read the instructions provided with whatever glucose monitoring device is available (Figure 6.5). Your instructor will point out several important steps to attain an accurate measurement. This device actually is a special-function spectrophotometer.

2. Before actually taking the blood sample to analyze, record when and what food was eaten at the last meal. This will be compared with the results obtained.

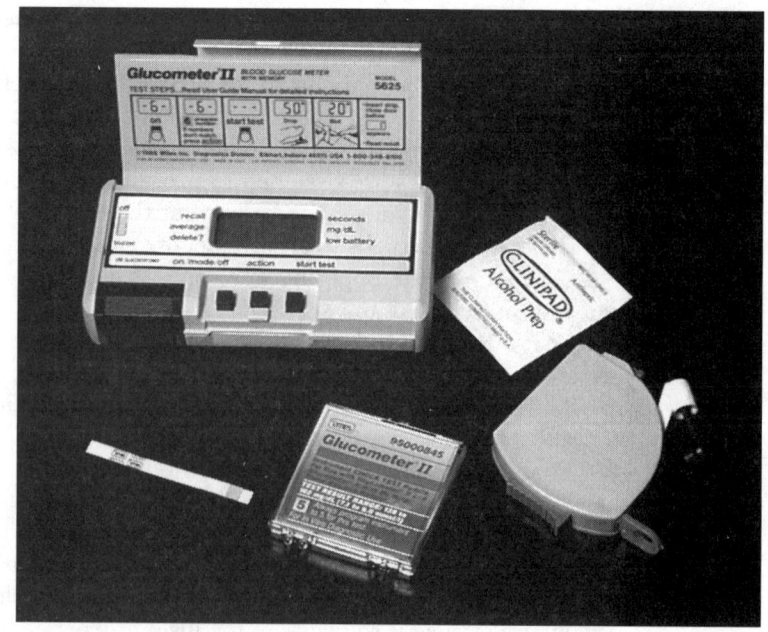

Figure 6.5
A Pocket Glucose Monitoring Device.
Pictured here is one of many inexpensive, easy-to-use devices that allow persons to monitor their blood glucose levels. This particular meter is the Glucometer II (Miles, Inc., Diagnostics Division, Elkhart, IN 46515).

3. Carefully wash the hands, and select a finger to be punctured. Clean the finger with an alcohol pad, and let the finger dry completely. Then, puncture the finger with the lancing device.

4. Collect an adequate-size drop, per instructions, on the chemically treated portion of the test strip. Blot the blood at the proper time with cotton or tissue as instructed.

5. Insert the test strip into the monitoring device chamber, close the door, and wait for the reading. The instrument used probably comes with normal ranges for blood sugar; howeve, a range of 70 to 110 mg per 100 ml is considered to be normal for a fasting glucose value. Unless a student has diabetes, everyone in the class should fall into this range.

6. Compare the information concerning meals (recorded earlier) with the actual glucose levels in the blood. The variety of food items eaten and the variable times since eating should further illustrate the careful control that homeostatic feedback mechanisms can provide.

QUESTION: To determine if the body's glucose regulatory control system is working properly, a glucose tolerance test can be done. The test involves the subject drinking a very sweet, flavored drink and then measuring the blood and urine sugar levels at timed intervals. What hormones are involved in the regulation of blood sugar? In other words, what hormone regulatory system and glands are being checked with this test?

Observations:

Conclusion:

The Gonads

The **gonads** (*gone* (G), offspring) belong to two systems of the body. In this case, the gonads are truly endocrine glands, because both (i.e., testes in men, ovaries in women) produce hormones to regulate the reproductive system. They also produce the cells that are responsible for allowing reproduction. In women, the gonad is the **ovary,** and in addition to the eggs, or **ova** (singular, *ovum*), that it produces, it releases **estrogen** and **progesterone.** In men, the gonad is the **testis** (plural, *testes*), which produces sperm and **testosterone.** The gonads are the targets for at least two hormones produced by the pituitary gland, FSH and LH. FSH, or follicle-stimulating hormone, is named for its function in women regarding the group of cells that help to produce the eggs. LH, or luteinizing hormone, also is named for its role in women; LH stimulates the follicle, once it ovulates or sheds its egg, to become a specialized group of endocrine cells called the **corpus luteum** (*corpus* (L), a body; *lute* (L), yellowish).

QUESTION: As stated, both FSH and LH are named for their role in the female body. Considering they perform a comparable task in men, what do you think these hormones do in the male body?

Observations:

Conclusion:

Journal Note Body builders can develop their muscles to massive sizes. Look at a body-building magazine, and note what kinds of supplements are advertised. What kinds of hormones will stimulate development of these massive muscles? What are the side effects of these chemicals? Do you think these "over the counter" chemicals really work? Did you know that there are associations and magazines that focus on drug-free body building?

Section 7

The Nervous System—Part 1

Note to the Student

Our study of the nervous system is divided into two sections in this manual. Even an introductory coverage of the nervous system is so extensive, it is best to break this large topic into separate areas.

The nervous system is the most complicated system you will learn; however, a convenient organizational structure should assist you in understanding its anatomy and physiology. The system usually is divided into the central nervous system (CNS) and the peripheral nervous system (PNS). Table 7.1 describes the structural and functional organization of the nervous system. Most of these parts are referred to in one of the two sections on the nervous system; some aspects of sensory and reflex physiology are explored as well.

Journal Note The only means of communication in the nervous system is the action potential, and a good analogy of action potential is electrical current flowing through a wire. Just as action potentials allow the body to do all types of activities because of differences in receptors and effectors, think of all the different things we can do with electricity. List the many different uses for electricity in our houses, buildings, and manufacturing facilities. This should help you to better understand action potentials.

Architecture of the Nervous System

The structural and functional unit of the nervous system is the **neuron,** or nerve cell (Figure 7.1). For ease of naming and labelling, we will describe a "typical" nerve cell. The neuron consists of a **cell body,** which also is referred to as the **soma** *(soma* (G), a body) or **perikaryon** *(peri* (G), around; *kary* (G), a nucleus). The cell body contains all of the organelles typically found in cells (review the cytology and tissues sections). Extending from the periphery of the cell body are **dendrites** *(dendr* (G), a tree; *ites* (G), belonging to, having to do with), which are cytoplasmic extensions bounded by cell membrane that carry signal information to the cell body and the rest of the cell. They also increase the input surface area of the nerve cell.

Table 7.1   Organization of the Nervous System

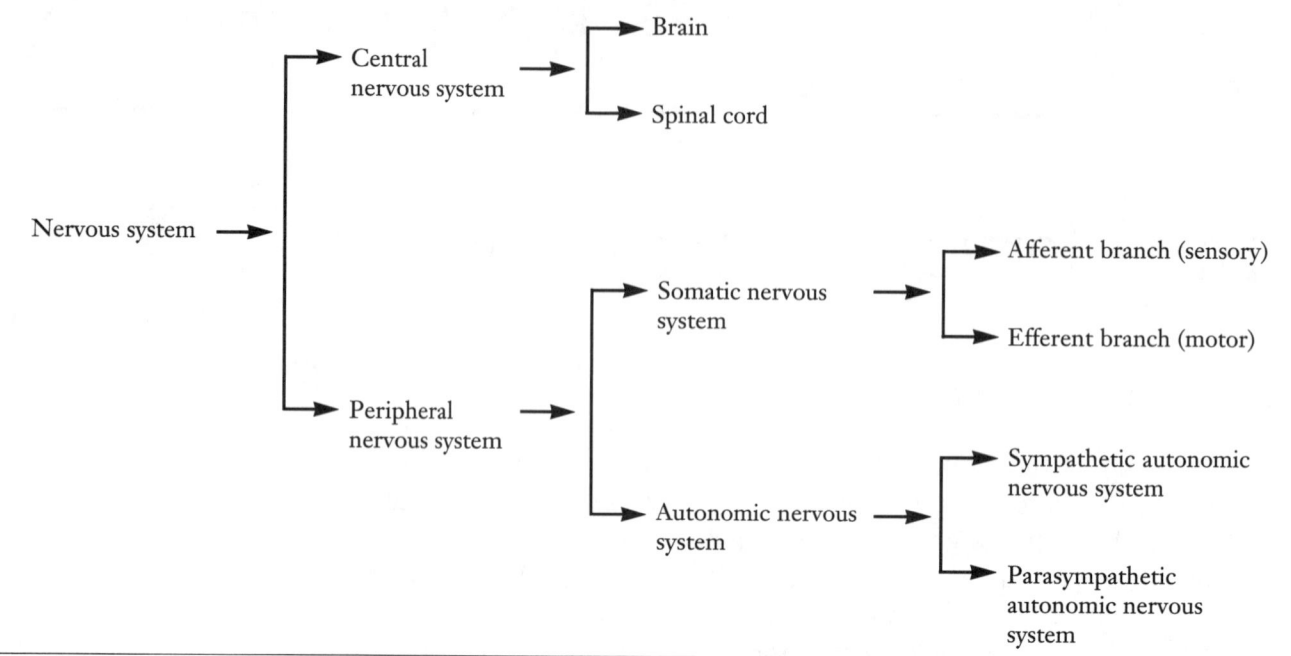

Usually there is a single, larger extension of membrane-bound cytoplasm, which is the **axon** (*axon* (G), an axis). This is the process that carries information away from the cell body, away from the input area of the cell, and toward the next cell. Therefore, there is a directionality in nervous conduction, from dendrites to cell body to axon to the next cell (if the signal is strong enough). The axon of a "typical" cell often is surrounded by specialized cells, which are called **Schwann cells,** that provide an insulation layer around the axon. The insulation's main component is myelin and, therefore, is referred to as the **myelin sheath.** The axon may be several feet long, as it is in the arms and legs, where sensory neurons run from the fingers or toes to the spinal cord, and from the spinal cord to the fingers and toes in the case of motor neurons. The insulation is truly an electrical insulation, but its physiological importance is to speed signal transmission rather than simply maintain signal separation.

Anatomy of the Spinal Cord

One's first guess might be that study of the CNS would start with the brain. The brain is such a complicated entity, however, that some basic terminology can be covered by discussing the spinal cord first.

The **spinal cord** is located in and protected by the vertebral canal, which is composed of the individual arches of the vertebrae. The spinal cord begins at the base of the brain, at the medulla, and terminates at L1–L2 (i.e., the level of the first and second lumbar vertebrae). In cross-section, there is a characteristic outer area containing white or myelinated matter (neurons) and an inner area containing gray or unmyelinated matter (also neurons), and gray matter typically is in a butterfly configuration. An important point about spinal cord organization is that the gray matter is where information processing occurs and the white matter is designed to carry information from one part of the spinal cord to another part of the CNS.

Figure 7.1
A Typical Nerve Cell.
This figure illustrates the three typical components of a nerve cell or neuron: the dendrites, the soma or cell body, and the axon. Also shown are the endings of the axon.

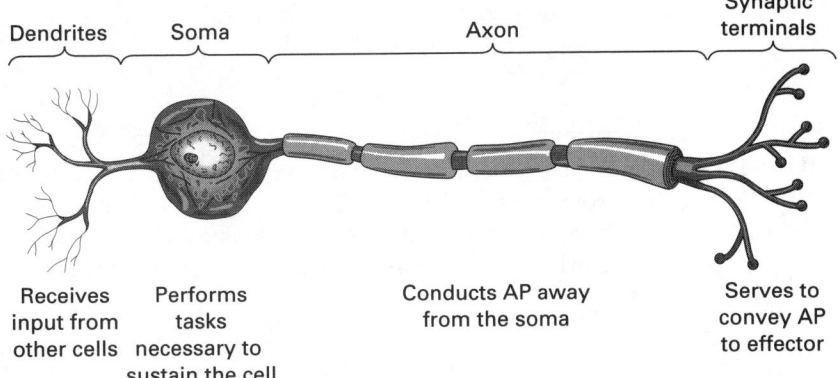

Cross-Section of the Spinal Cord

Objectives:

- To appreciate the types of matter that make up the spinal cord.
- To understand the regions of the spinal cord so that to structures in these regions can be located.

Materials:

- Microscope.
- Spinal cord slide, cross-section.

Procedure:

1. Obtain a slide of the spinal cord, and be aware of the stain used for the slide you are observing. Several commonly used stains are applied to sections of nervous tissue to show certain types of cells. Your slide probably is stained with hematoxylin and eosin, which cause the cell nucleus to be visible. Silver staining, which is commonly used with nervous tissue, makes the extensions of the cell visible.

2. Refer to Figure 7.2, and identify the structures or areas indicated.

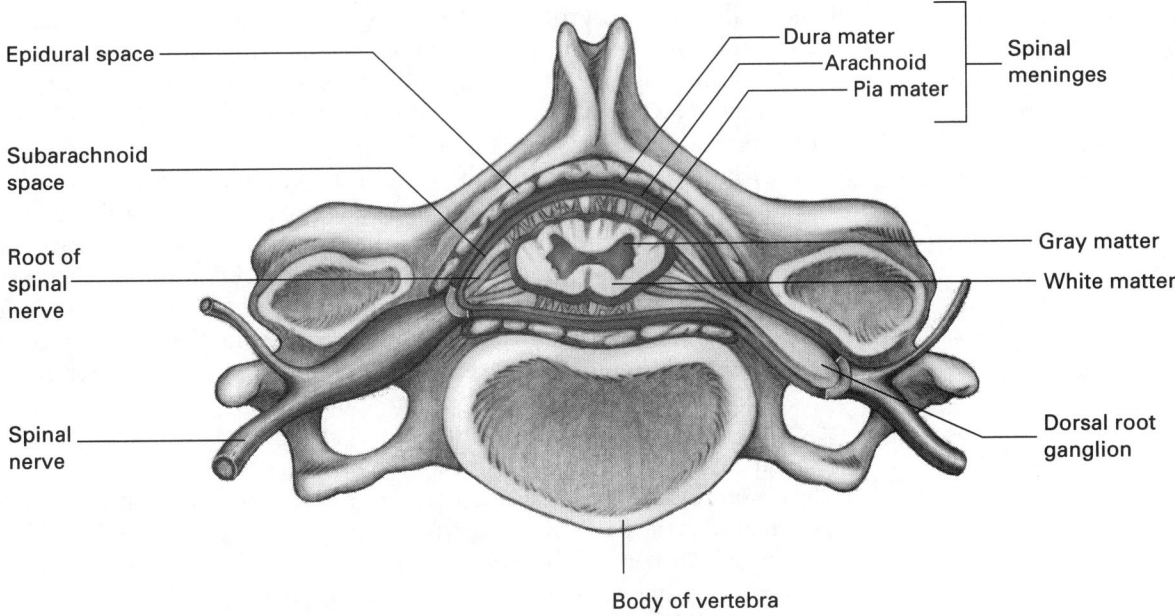

Figure 7.2
The Spinal Cord.
This cross-section of the spinal cord shows the major regions of the cord, the gray matter, and the white matter. Also shown are the major anatomical terms associated with the cord.

Anatomy of the Spinal Cord

3. Considering the texture and consistency of nervous tissue, the section may not appear as illustrated in Figure 7.2 because of distortion of the tissue caused during sectioning.

QUESTION: As mentioned, white matter contains myelinated fibers and brings information to and from the brain. Gray matter is unmyelinated and serves as processing areas for that information. Based on the appearance of the spinal cord section and this information, in which area of the spinal cord, white or gray, are the cells cut in cross-section?

Observations:

Conclusion:

Everyone is familiar with the term "nerve," but not everyone may be sure how it is defined in anatomy and physiology. In the arrangement of the nervous system, axons for the similarly functioning neurons typically are bundled together as nerves, and the cell bodies for these same neurons also are grouped in the same location. This means that cell bodies and axons are not intermixed throughout the nervous system. Table 7.2 lists terms related to this structural characteristic.

There are areas of gray matter in the brain and spinal cord that contain cell bodies for a large number of cells all involved in a specific monitoring or control function. In the CNS, this type of grouping is called a **nucleus;** in the PNS, this is called a **ganglion** (plural, *ganglia*) (*ganglion* (G), a swelling). Refer to Table 7.1 for the structure of the PNS. Whereas a nerve is truly defined as a bundle of neurons in the PNS, it is only the axons, functionally speaking, that are bundled together to form a nerve. In the CNS, this same grouping of axons is called a **tract.** The white matter of the spinal cord and brain contains tracts, and the gray matter of these structures contains nuclei.

Considering the entire length of the spinal cord, two enlarged areas are quite apparent. The one formed by the cervical vertebrae (i.e., C4 to T1) is referred to as the **cervical enlargement** (Figure 7.3). The function of the cervical enlargement is the operation of the upper limb, or the arm. The spinal cord narrows only to enlarge again, at T9–T12, as the **lumbar enlargement,** which controls the operation of the leg. Because of all the muscle control and sensory input from the arms and legs, these are busy areas of the cord. This often is a point of confusion. The cervical enlargement is located in the cervical region of the vertebral column, so why is the lumbar enlargement in the thoracic part of the vertebral column? Why not call it the thoracic enlargement? (Refer to your text for the differences in the configuration of the segments of the vertebral column and the spinal cord.)

If a longitudinal section of the spinal cord were made, the gray matter areas in the spinal cord, while continuous, would look like beads on a string. There would be 31 beads on this string, because there are 31 **spinal nerves** coming off the spinal cord. Each bead is the base for the nervous activity entering the cord from that spinal nerve or exiting the cord to that spinal nerve. Remember that the spinal cord ends at L1–2 but still contains 31 segments, one for each spinal nerve. Therefore, the 31 segments of the cord do not correlate with the same areas of the vertebral column, so the lumbar enlargement truly is the enlarged lumbar areas or segments of the spinal cord,

Table 7.2 Terminology Related to Similar Structures in the CNS versus the PNS

	CNS	PNS
Group of nerve cell bodies	Nucleus	Ganglion
Bundle of neurons	Tract	Nerve

despite their positioning in the thoracic region of the vertebral column. Each spinal nerve comes off the cord as a **dorsal** and a **ventral root.** (We will see the functional significance of this when we discuss reflexes.)

The spinal cord ends before the vertebral canal ends, but this does not mean the rest of the canal is empty. The spinal nerves may come off the cord more superiorly, but they continue down the vertebral canal until they reach the vertebral column point, where they should exit or enter the canal. Thus, the lower lumbar and sacral portions of the vertebral canal contain the spinal nerves for segments L3 to S5 (C0). This is called the **cauda equina** (*cauda* (L), the tail; *equin* (L), pertaining to horses), because it looks like a horse tail. Running from the actual end of the spinal cord, through the middle of the cauda equina, and anchoring the spinal cord to the end of the vertebral canal is the **filum terminale** (*filum* (L), a thread; *termin* (L), an end). This is a connective tissue fiber that extends from the meninges, which is the connective tissue covering of the spinal cord.

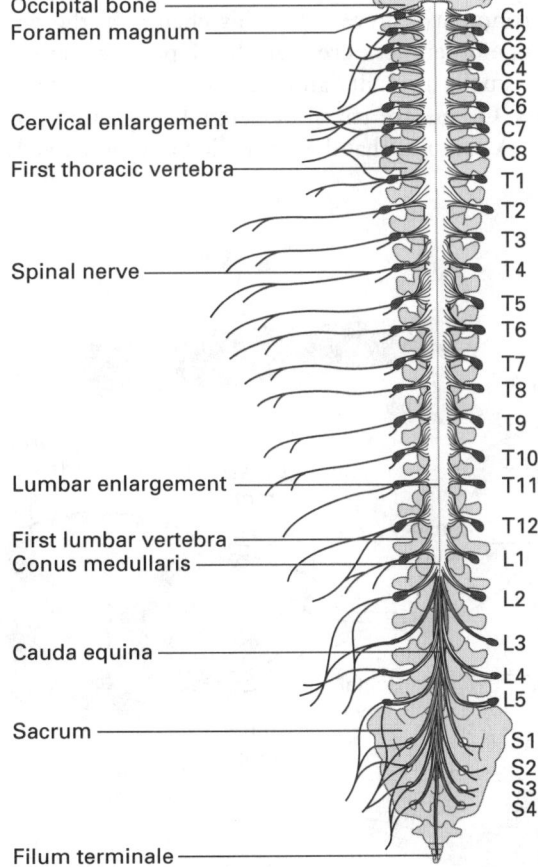

Figure 7.3
The Spinal Cord.
This posterior view shows the spinal cord with the vertebral arches removed. Note the spinal cord ends before the end of the vertebral column, at about the level of the first or second lumbar vertebrae. The spinal nerves, however, continue down the vertebral canal and exit at the appropriate level.

Anatomy of the Spinal Cord

The Meninges

The brain and spinal cord are surrounded and protected by a set of thin, connective tissue layers, which collectively are called the **meninges** (*mening* (G), a membrane). There are three layers to the meninges. The outermost, or the layer furthest away from the spinal cord, is the **dura mater** (*dura* (L), hard; *mater* (L), a mother). It is dense, connective tissue that provides a tough, protective tube for the spinal cord. It also surrounds the brain. The space between the dura and the vertebral canal is referred to as the **epidural space,** and this is where anesthetics are injected for anesthesia of the lower body. The space below the dura is the **subdural space,** and the next layer of the meninges is the **arachnoid** (*arachna* (G), a spider web) because of the delicate, web-like arrangement of the connective tissue. The **subarachnoid** space is where the **cerebrospinal fluid** flows around the spinal cord and the brain to protect and to cushion them. The innermost, and the thinnest, layer of delicate connective tissue fibers and blood vessels is the **pia mater** (*pi* (L), delicate, tender), which is connected directly to the spinal cord and brain.

Much of the discussion so far has been about the anatomy of the nervous system. What about the functioning of the spinal cord?

Physiology of the Spinal Cord: Reflexes

Now, let us consider one of the principal activities or functions of the spinal cord: as a center for reflex activity. A **reflex** is an automatic response of the nervous system. The first component of a **reflex arc** is a **receptor** that tells the system of some change in the environment (Figure 7.4). This change in the environment is called a **stimulus.** Any change in temperature, light level, pressure on the body, sound or the noise level, types of smells present, and so on represent various stimuli. Many stimuli elicit responses. Others are trivial or below background noise, however, and do not elicit a response. A stimulus that does not elicit a response is described as being **subthreshold.**

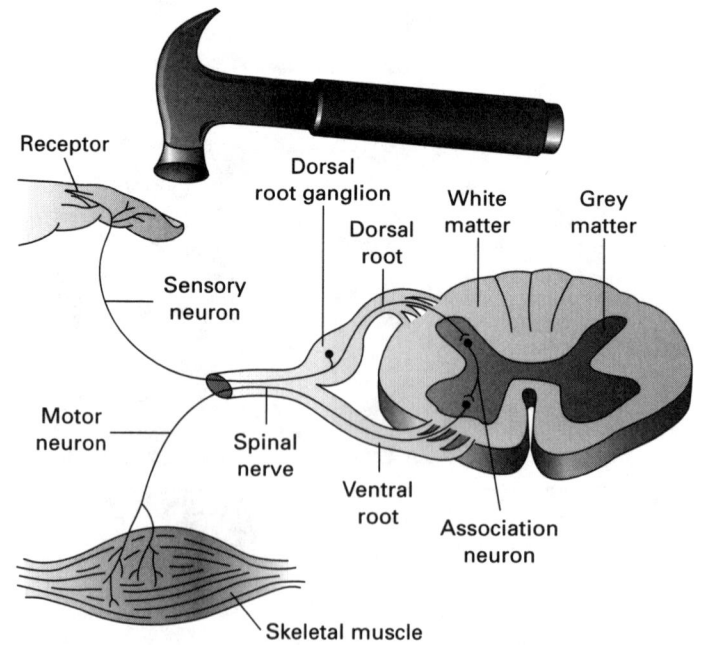

Figure 7.4
A Reflex Arc.
There are five components to a typical reflex arc: a receptor, a sensory neuron, an association neuron, a motor neuron, and an effector. Reflexes occur without immediate awareness by the brain. The brain learns of the reflex activity either via branches of the interneuron continuing to the brain or through visual information arriving at the brain.

A receptor is a type of nervous structure that detects changes in the environment, either inside the body (i.e., the internal environment) or outside the body (i.e., the external environment). A reflex arc also contains a **sensory neuron** that connects the receptor with the spinal cord. On approaching the spinal cord, the spinal nerves split into a dorsal and a ventral root. The dorsal root contains all of the sensory neurons, and the **dorsal root ganglion** contains the cell bodies for all of these neurons. Axons of the sensory neurons enter the spinal cord and terminate in the posterior gray horn, where they synapse with the next component of a reflex. Typically, there is an **interneuron,** or an **association neuron,** in the gray matter of the cord that connects the sensory neuron with the **motor neuron.** The motor neuron's cell body is located in the anterior gray horn; its axon exits the cord in the ventral root, and the ventral root fuses with the dorsal root to form a spinal nerve. The final component of the reflex arc is an **effector,** which is a muscle or gland that carries out the action associated with the proscribed reflex.

QUESTION: Did you ever realize that only a muscle or gland can actually "do" anything in the body? Other tissue types in the body perform important tasks, but they do not carry out responses to stimuli. Name as many of the various types of actions that a muscle and gland can "do" for the body as you can.

Observations:

Conclusion:

The specific effector that performs most of the conscious responses we think of as reflex actions are the skeletal muscles. Smooth muscle, cardiac muscle, and glands are the effectors of subconscious reflexes. Note there is no required connection of the reflex arc to the brain; therefore, the brain is not involved in the loop that operates the reflex action. When (and if) the brain finds out about the reflex action, it is through other senses observing the action or by a direct signal to the brain generated by a branch of the interneuron, whose signal arrives at the brain after the reflex action is complete or underway.

A reflex that results when the stimulus signal arrives at a sensory fiber on one side of the spinal cord and the motor signal exits on the same side of the cord is called an **ipsilateral reflex.** There also may be bits of information about this reflex that must be transmitted to additional segments of the cord, thus making this an **intersegmental reflex.** The simplest reflexes do not include the interneuron as part of the circuitry. As a result, there is but a single synapse in the gray matter of the cord, thereby making this a **monosynaptic reflex.** Most reflexes, however, spread to surrounding areas and have an interneuron; these are referred to as **polysynaptic reflexes.** For certain reflexes, it is necessary for information received on one side of the cord to require a response on the opposite side, and this type is called a **contralateral reflex.**

Reflexes provide clinicians, and specifically neurologists, with a means of detecting damage to the spinal cord during a neurological exam. As reflexes that occur in various parts of the body are specific to precise locations in the spinal cord, their absence or deterioration indicates damage to or degeneration of that specific site.

Reflex Demonstration

To demonstrate a reflex requires the appropriate stimulus to be delivered at the right point on the body surface. Follow the directions provided, and use the figures to help you demonstrate these important reflexes.

Patellar or Knee-Jerk Reflex

The best known and most common example of a stretch reflex is the patellar or knee-jerk reflex. Stretch reflexes help to maintain tone in all the body's muscles so they are ready for movement whenever needed.

Knee-Jerk Reflex

Objectives:

- To demonstrate the knee-jerk reflex as an example of a typical reflex.
- To show a principal role of the spinal cord as a center for reflex activity.

Materials:

- Volunteers.
- Percussion or reflex hammer.

Procedure:

1. To elicit the knee-jerk reflex, strike the patellar ligament with a percussion hammer. Several attempts may be needed to find the best striking location and pressure.

2. The volunteer will need to be sitting down, with his or her legs hanging free or crossed with the leg to be tested on top. Strike the ligament below the kneecap, and observe the response (Figure 7.5).

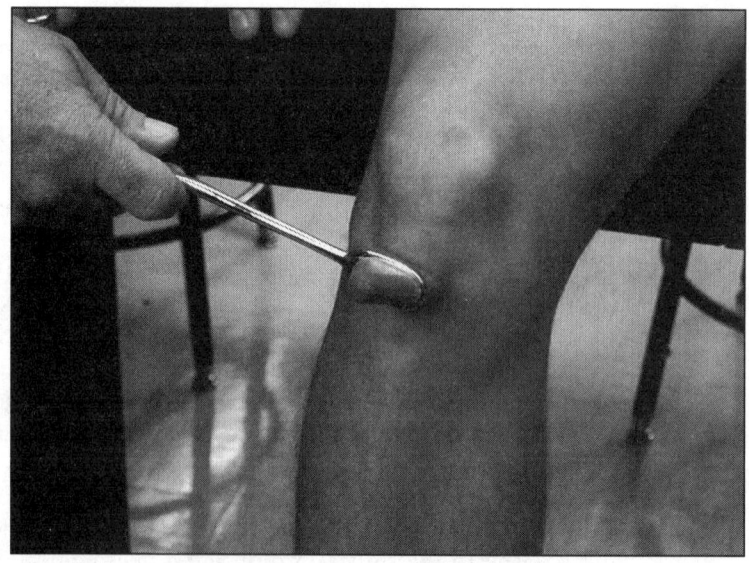

Figure 7.5
The Patellar Reflex.
Note the area in which the patellar ligament should be struck.

106 SECTION 7 The Nervous System—Part 1

QUESTION: What muscle or muscles are involved in the knee-jerk reflex? What are the receptors for this reflex?

Observations:

Conclusion:

3. In some persons, there is only a weak reflexive response at best. To increase the response, instruct the person to grip the fingers of one hand with those of the other as shown in Figure 7.6.

4. Now, tell the subject to pull the hands apart as hard as possible while maintaining the tension only in the upper body. **Do not tense the legs and lower body.** Demonstrate the knee-jerk reflex again. What happened?

QUESTION: The only signal that is carried over distances in the nervous system is the action potential. To start an action potential in a neuron requires the stimulus to reach and exceed threshold. Can you explain the increased response in the knee-jerk reflex just seen based on threshold for action potential firing?

Observations:

Conclusion:

Journal Note The next time you are at an athletic event, watch the participants and imagine all of the automatic movements they make as they go through the motions of their sport. Basketball is good for this type of activity, because spectators often are rather close to the participants. Alternatively, you can observe your bare feet and ankles as you stand in place and rock back and forth. Seemingly unconscious changes occur in muscle tension as the body's center of gravity shifts. Become an observer of people's movements as they perform their daily activities. Many reflexes are involved in most of what we do on a regular basis.

Figure 7.6
Procedure to Modify the Extent of the Patellar Reflex. With the legs relaxed and unrestricted, clench the fist as shown and pull forcefully. Do not pull so forcefully with the entire body, however, that the legs stiffen and are not relaxed.

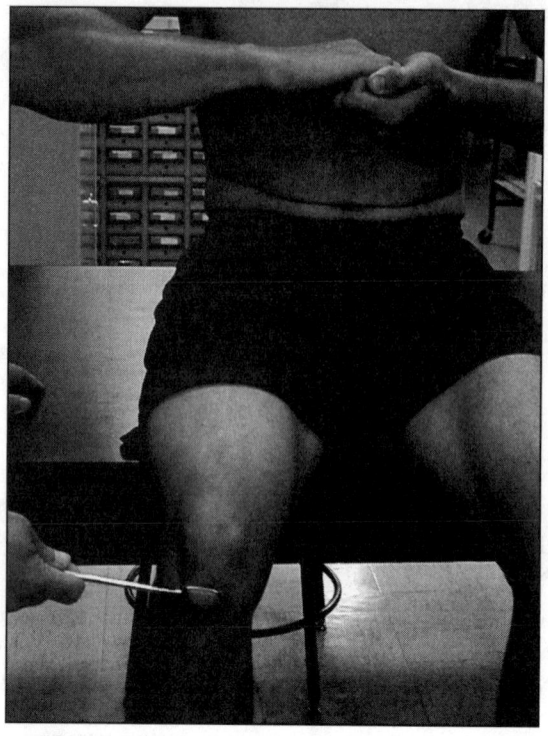

Ankle-Jerk or Achilles Reflex

Another stretch reflex that can be demonstrated is the ankle-jerk or Achilles reflex.

Achilles Reflex

Objectives:

- To demonstrate the Achilles reflex as an example of a stretch reflex.
- To show that reflexes can be demonstrated in several body areas.

Materials:

- Volunteers.
- Percussion or reflex hammer.

Procedure:

1. Have the subject remove his or her shoes and socks and to kneel on a chair. The subject's feet should extend off the front of the chair (Figure 7.7).

2. Strike the Achilles tendon at the level of the medial or lateral malleolus. The foot should exhibit what is referred to as a **plantar extension.**

Figure 7.7
Achilles Reflex.
Note the positioning to demonstrate the Achilles reflex.

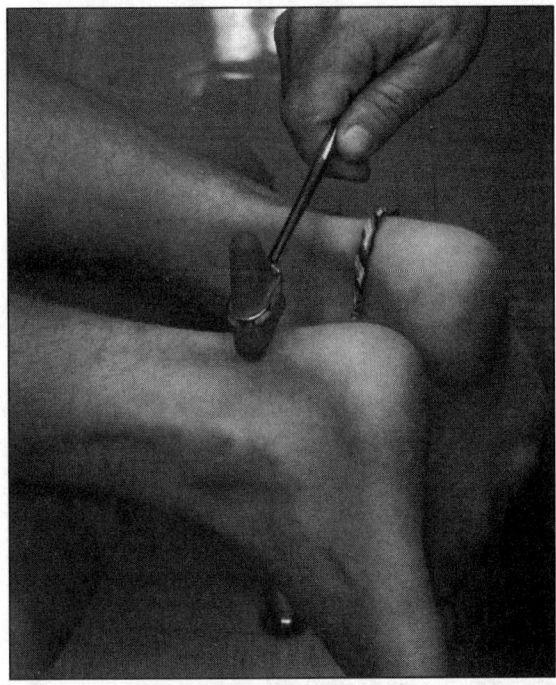

QUESTION: What types of sports activities would use this reflex frequently? (Hint: Think of activities that require jumping.)

Observations:

Conclusion:

Plantar and Babinski Reflexes

Unlike the previous two examples, the plantar and Babinski reflexes are not stretch reflexes. The plantar reflex is a response of the foot to stimulation with a blunt object from the lateral border of the foot curving across to the base of the big toe. This is an important reflex to test for damage to the lower spinal cord. The normal adult reflex is the plantar reflex, and the abnormal response in adults (but the normal response in infants) is the Babinski reflex or sign.

Plantar and Babinski Reflex

Objectives:
- To demonstrate the plantar and Babinski reflexes as examples of non-stretch reflexes.

- To appreciate the clinical significance of reflexes as a means to ascertain the integrity of the nervous system.

Materials:

- Volunteers.
- Percussion or reflex hammer.

Procedure:

1. Have the subject lie down on his or her back and remove the shoes and socks. To test a foot, have the subject cross his or her legs, with the test foot projecting out so that the foot's lateral border faces the ground.

2. Use a blunt object (i.e., the handle of the percussion hammer) and firmly, but not painfully, stroke the lateral border of the foot, curving with the base of the toes and ending at the big toe (Figure 7.8).

3. The normal adult response is a plantar flexion of the toes. This means that the toes align, become pointed, and bend toward the plantar surface of the foot. The abnormal adult response and normal infant response is called the Babinski reflex or sign, which is a dorsiflexion of the big toe and fanning of the other toes.

Figure 7.8
Plantar Reflex.
Note the positioning of the foot to demonstrate the plantar reflex. (a) Start with the handle of the probe on the lateral margin of the foot. (b) Proceed, firmly moving the probe along the lateral margin of the foot to the base of the toes and across to the big toe.

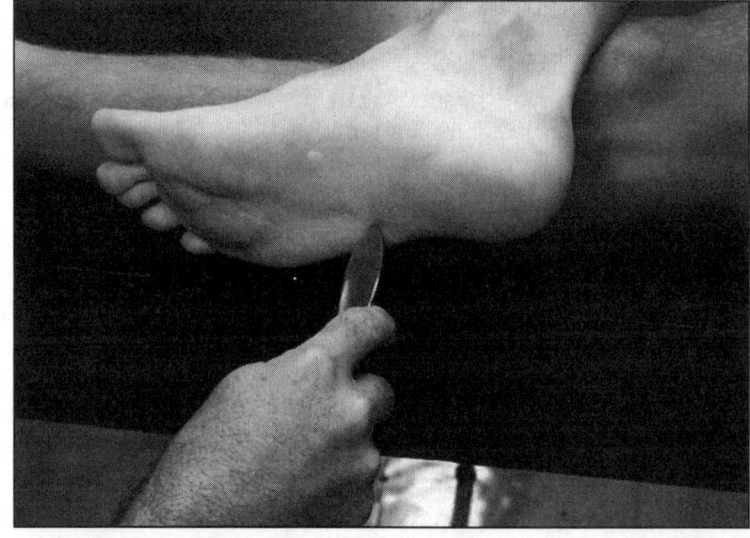

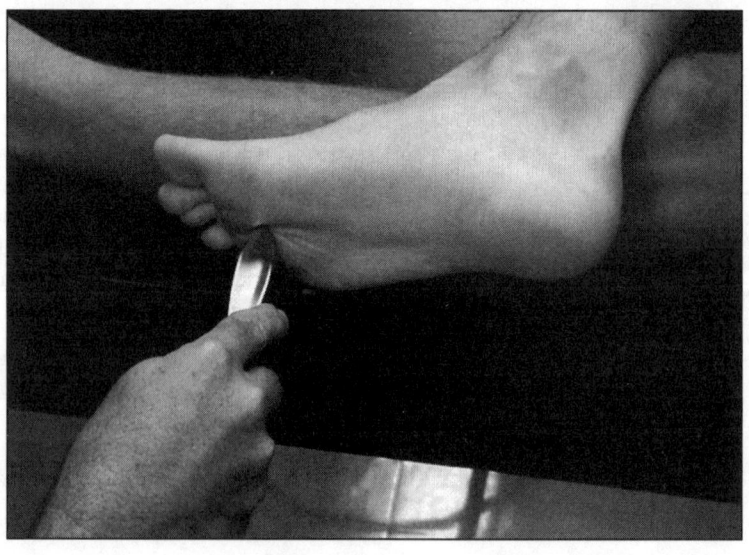

Figure 7.9
Plantar versus Babinski Reflex. The normal adult response as the bottom of the foot is stroked is referred to as the plantar reflex, which consists of a downward flexion of the toes. The Babinski reflex or sign consists of a dorsal flexion of the big toe and a lateral fanning of the other toes; in adults, this sign indicates damage to part of the brain or spinal cord.

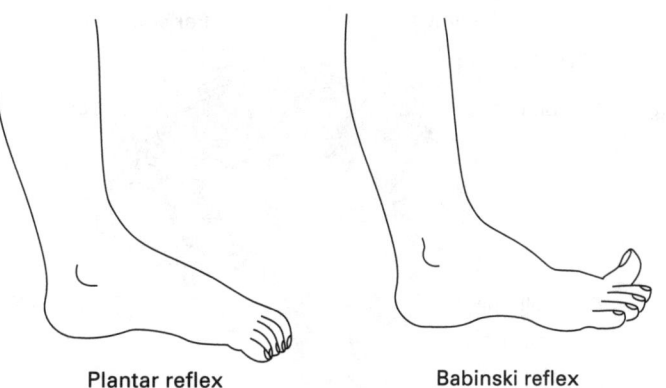

Plantar reflex   Babinski reflex

QUESTION: An abnormal plantar reflex or a Babinski sign indicates a damaged pyramidal or corticospinal tract (Figure 7.9). Where does this tract come from, where does it go, and what type of information does it carry?

Observations:

Conclusion:

Anatomy of the Brain

The brain typically is divided into four major regions based on different functions. These include:

1. The **brain stem**, which is further divided into the medulla oblongata, the pons varolii, and the midbrain;

2. The **diencephalon**, which consists of the thalamus and the hypothalamus;

3. The **cerebrum**; and

4. The **cerebellum**.

We now consider each of these parts, with at least some mention of the major activities they control.

The Brain Stem

The Medulla Oblongata

The **medulla oblongata** is a part of the brain that is approximately 2.5 cm in length and connects the entire brain with the spinal cord (Figure 7.10). All of the motor and sensory information that flows to or from the spinal cord must go through the medulla. There are many tracts that go through the medulla, and there are also several reflex

Figure 7.10
The Brain, Sagittal Section.
This sagittal section of the brain shows its major regions.

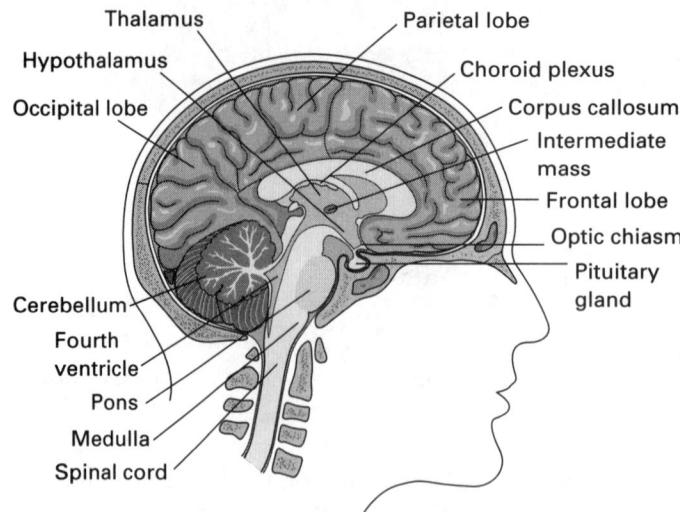

centers. The medullary rhythmicity center controls breathing, the cardiac inhibitory center constantly tries to slow the heart, and the vasomotor center maintains blood pressure. The medulla oblongata is described as a primitive part of the brain. In this sense, "primitive" means that the medulla is similar to that found in lower-vertebrate animals, because all animals need to control these basic functions.

The Pons

The **pons** (*pons* (L), a bridge) carries the information from the medulla to the rest of the brain (Figure 7.10). It also has additional control centers that are involved in regulating respiration. These centers are the **apneustic** and **pneumotaxic centers.**

The Midbrain

Two very important reflex centers in the midbrain are the **superior** and **inferior colliculi** (*collicul* (L), a little hill) (Figure 7.10). These are a pair of bumps on the posterior side of the midbrain, and they are hidden from view by the cerebellum. The superior colliculi cause us to automatically move the head and eyes in the direction of some visual cue. The inferior colliculi cause us to automatically rotate the head and trunk when we hear a sound. The purpose of both is to allow us to take in as much visual and auditory information as rapidly as possible when alerted by visual and auditory stimuli.

The Reticular Formation

The **reticular formation** is an area of the brain that actually begins in the spinal cord and extends through the medulla, pons, midbrain, and diencephalon. It controls the flow of sensory information to the brain. The functioning of the **reticular activating system,** as this area also is called, is realized as we doze off to sleep.

QUESTION: If you have ever been jolted to alertness just after falling asleep, you have experienced the reticular activating system failing to do its job. Can you explain what might have caused this sudden change in your level of consciousness?

Observations:

Conclusion:

The Cerebellar Peduncles

There are connections between the medulla, pons, midbrain, and the cerebellum, and the structures that connect these areas are called the **peduncles** (*peduncul* (L), a little foot). The superior peduncle connects the midbrain with thecerebellum, and the middle and inferior peduncles, respectively, connect the cerebellum with the pons and the medulla. There is both a flow of sensory information into the cerebellum and of motor information out through these connections.

The Diencephalon

Diencephalon is a name that is used when describing one of the developing regions of the embryonic brain. It is at the base of the brain and very centrally located, being surrounded by the cerebrum on most of its sides. It contains the **thalamus** (*thalamus* (G), a chamber, inner room) and the **hypothalamus** (*hypo* (G), under, beneath) (Figure 7.10). The thalamus is a relay station for sensory information (except olfactory) coming from the spinal cord and sense organs, and also for motor information from the brain to the spinal cord. The hypothalamus is located below the thalamus, and because of its connection through a funnel-shaped stalk called the **infundibulum** (*infundibulum* (L), a funnel), it is the neuroendocrine connection to the pituitary gland. The hypothalamus and pituitary gland are very involved with homeostasis. Many important primitive functions, such as feeding, thirst, body temperature regulation, and reproduction, are controlled by nuclei in the hypothalamus and by the pituitary.

The Cerebrum

The cerebrum raccounts for most of the brain, and it is the location of all higher-level functioning. It is divided into two halves, the **left** and **right cerebral hemispheres,** which are separated from each other by the **longitudinal fissure.** A fissure is a deep groove between sections of the brain (Figure 7.11). There also are many shallow grooves and bumps in between the grooves, which gives the brain its characteristic appearance. The grooves all over the surface are called **sulci** (singular, *sulcus*) (*sulcus*

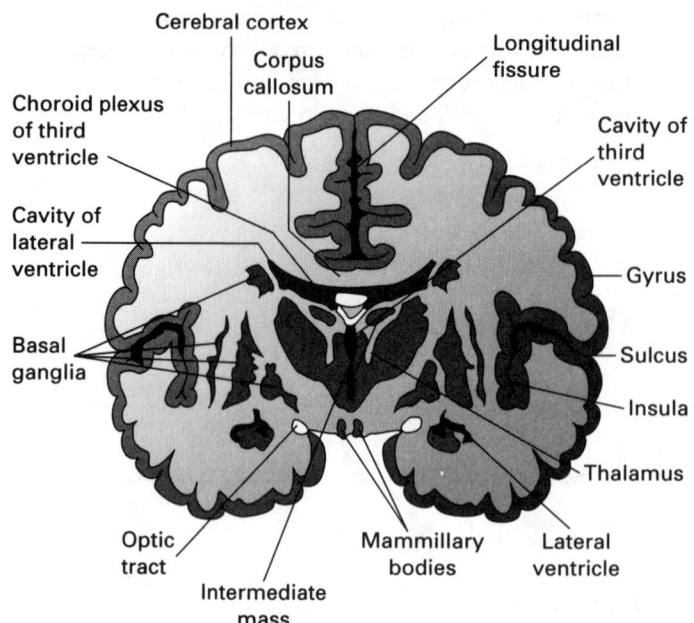

Figure 7.11
The Brain, Frontal Section.
This frontal or coronal section of the brain allows us the see both sides, as opposed to the section in Figure 7.10, which only shows half of the brain.

(L), a groove), and the bumps are called **gyri** (singular, *gyrus*) (*gyr* (L), round). Remember that the meninges surrounding the spinal cord also cover and protect the brain, and the thin, blade-like extension of the dura mater between the two hemispheres is referred to as the **falx cerebri** (*falx* (L), a sickle). Each hemisphere is subdivided into lobes that bear the same names and locations as the bones of the skull over these areas. There are a pair of **frontal, parietal, occipital,** (Figure 7.10) and **temporal lobes,** or one of each pair in each hemisphere. The two hemispheres are connected to each other by a median structure called the **corpus callosum** (*corpu* (L), a body; *callo* (L), hardened). This structure contains fibers that connect the two halves of the brain so that one side knows what the other is doing. The outermost surface, gray matter layer of the cerebrum is referred to as the **cerebral cortex** (*cortex* (L), bark, shell), contains billions of cells with many interconnections, and represents the processing area for all higher-level thought processes.

Within the cerebral hemispheres are two cavities, or **ventricles** (Figure 7.11), which contain a tissue that produces cerebrospinal fluid. Thus, the two lateral ventricles are filled with cerebrospinal fluid, and they are continuous with other ventricles and the subarachnoid space within the protective coverings of the brain and spinal cord. Cerebrospinal fluid provides a cushion to dissipate the force of impacts by dispersing the pressure into the fluid rather than into the brain tissue itself.

The Cerebellum

The cerebellum is shaped like a smaller version of the cerebrum (Figure 7.10). It has bumps, which are called **folia** (*folia* (L), a leaf) and grooves, which are called **sulci.** Its outermost layer also is gray matter, and it is called the **cortex.** The cerebellum coordinates motor control of muscle activity with the sensory information from the body. This function might be compared to that of an "automatic pilot," making sure the actions decided on by higher brain centers actually are performed. The deep groove running between the occipital lobe of the cerebrum and the cerebellum is called the transverse fissure.

The Cranial Nerves

There are 12 pairs of cranial nerves. They are referred to as **cranial nerves,** because they arise from brain structures rather than from the spinal cord itself (Figure 7.12). They are either sensory, motor, or mixed in function. Table 7.3 lists these nerves and the activities they control. The last column in the table is a list of words that together compose a mnemonic device to help you memorize the order of the nerves.

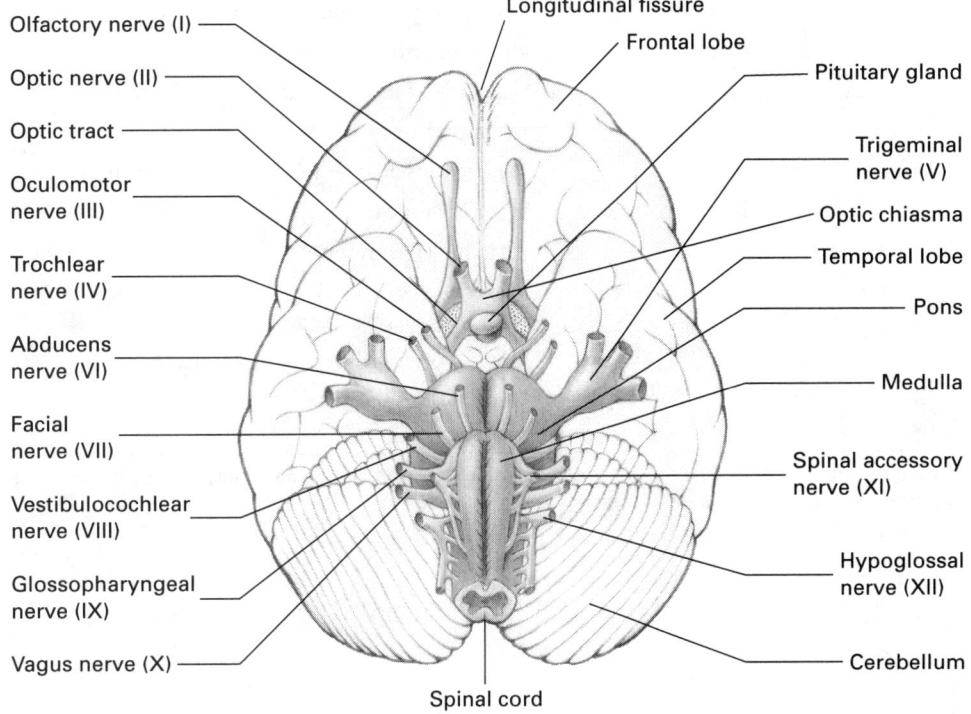

Figure 7.12
Base of Brain, Superior View. This superior view shows the base of the brain as removed from the skull, which allows us to identify the 12 cranial nerves where they leave the brain case.

Table 7.3 Cranial Nerves: Their Action and How to Remember Them

Cranial Nerve	Activity	Mnemonic Device
I. Olfactory	Sense of smell	On
II. Optic	the Sense of vision	Old
III. Oculomotor	Eye movement and light control	Olympus
IV. Trochlear	Eye movement	Towering
V. Trigeminal		Tops
Branch 1. Ophthalmic	Eyes, tear glands, upper eyelids, forehead	
Branch 2. Maxillary	Upper mouth and skin of face	
Branch 3. Mandibular	Scalp, lower mouth, and jaw muscles for chewing	
VI. Abducens	Eye movements and sensations	A
VII. Facial	Tongue sensations, facial muscles, tear and salivary glands	Finn
VIII. Vestibulocochlear (Auditory)	Inner ear for hearing and equilibrium	And*
IX. Glossopharyngeal	Taste, touch in mouth and throat, salivary glands	German
X. Vagus	Movements of speech and swallowing, heart rate, peristalsis, visceral sensations	Viewed
XI. Accessory	Muscles of mouth, pharynx, larynx, neck, and back	A+
XII. Hypoglossal	Muscles of the tongue	Hop

*The older name for the vestibulocochlear nerve was auditory, so that the mnemonic devise specifies an "A" rather than a "V."
+Some sources refer to the accessory nerves as the spinal accessory so that the "A" between "Viewed" and "Hops" is "Some."

Section 8

The Nervous System—Part 2

Note to the Student

After introducing the organization of the nervous system as well as the structure and function of the spinal cord and brain in the last section, we now concentrate on sensory structures and their function.

Sensory Physiology

There are two categories of senses: **general senses,** and **special senses.** General senses are distributed throughout the body and do not require a distinctive sensory-receptor structure to detect the sensation. Special senses are localized to specific body locations and have a distinctive, often sophisticated receptor apparatus involved in detection of stimuli. The special senses are those of taste, smell, vision, hearing, and equilibrium.

The Gustatory Sense

The gustatory (*gusta* (L), taste) sense more commonly is referred to as the sense of taste. The receptors are the taste buds, which are located on the tongue (associated with the fungiform and circumvallate papilla) and on the insides of the cheeks and into the pharynx. If one is available, examine a slide of a papilla, and observe the taste buds in the walls of the papilla (Figure 8.1). It is fairly well agreed there are four principal taste sensations: sweet, sour, salty, and bitter. The first three are associated with common foods, but bitter usually is a "warning" taste, which tells the individual the food is either poisonous or unsuitable for consumption (i.e., spoiled). There are thought to be four different types of buds, one for each of these tastes, and specific tastes are detected on different areas of the tongue (Figure 8.2). Sweet is detected on the tip of the tongue, sour on each side, and salty overlapping the anterior portion of the sour-detection area and the lateral parts of the sweet-detection area. Bitter is sensed on the posterior-superior area of the tongue (where gagging is triggered by a tongue depressor or finger). There is recent evidence that any taste can be detected anywhere on the tongue, but that the threshold for each taste varies. Other animals apparently can detect some other tastes, including water. Humans cannot detect water as a taste sensation.

Figure 8-1

Cross-section of a Fungiform Papilla on the Tongue and an Enlarged Taste Bud.
Note the taste buds in the spaces around the papilla. Both the fungiform and circumvallate papilla have taste buds associated with them. Note also the opening of the taste bud, which is the taste or gustatory pore. Taste buds are replaced by the basal cells in the taste buds every 10 to 15 days.

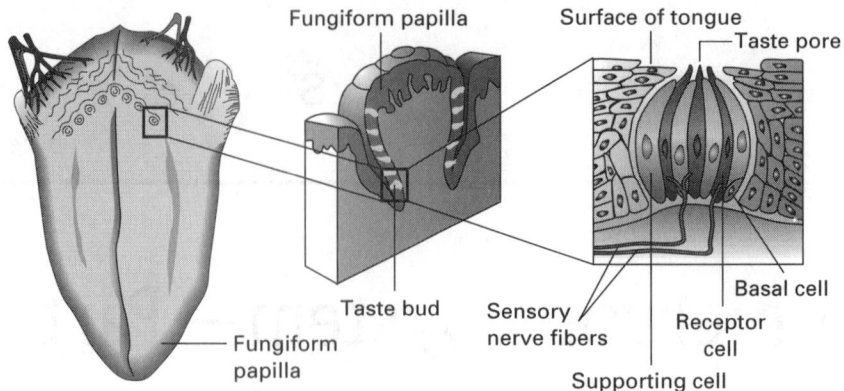

Taste Buds

Objectives:

- To understand the location of the taste buds in relation to the surface of the tongue and mouth.
- To appreciate the structure of a taste bud.

Materials:

- Microscope.
- Tongue surface slide, section showing fungiform papillae.

Procedure:

1. Obtain a slide of the tongue that contains a portion of the surface to show the fungiform papillae.

2. Place the slide on the microscope stage, focus on the surface area, and find the fungiform papillae. There are also finger-like papillae called filiform papillae, but these typically do not have taste buds associated with them.

3. Look at the lateral areas of the papillae for bulb-shaped structures: the taste buds. Refer to Figure 8.1 for details of their structure.

4. Look at several taste buds, and try to find an opening into the space around the papillae. This is called the gustatory pore.

Figure 8.2

A Map of the Tongue.
This map of the tongue shows the areas that are sensitive to the four tastes. Note that bitter is detected in the very back of the mouth, where gagging is produced by mechanical stimulation.

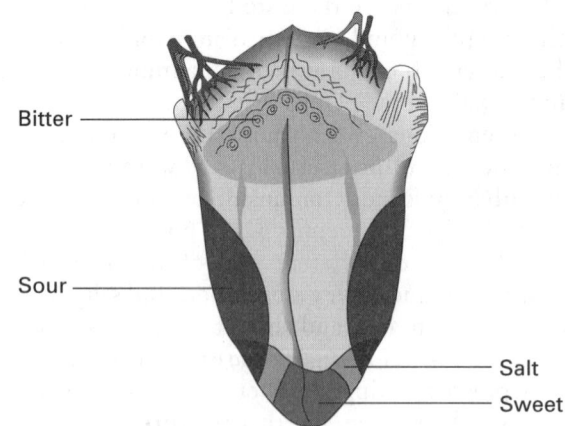

118   SECTION 8   The Nervous System—Part 2

QUESTION: The receptors for the sense of taste probably take as much—or more—abuse than any of the special senses. From your own experience, relate the types of materials or conditions that can damage or overstimulate taste receptors.

Observations:

Conclusion:

QUESTION: While discussing the four taste sensations, you may have been thinking about other sensations associated with food. Hot and spicy tastes are not included in the four sensations mentioned earlier. The chemical capsaicin seems cause the burning sensation of spicy foods. Which receptors do you think might be involved in this burning, painful sensation as we eat a jalapeño?

Observations:

Conclusion:

Demonstrating the Gustatory Sense

This test to demonstrate the gustatory sense, as well as most of the sensory physiology tests must be blind tests. This means there should be an experimenter and a subject to be tested. The persons being tested must be positioned in such a way, or instructed to close their eyes, so that they will not know what material is being tested. The experimenters also should randomize the procedure so that, for example, subjects will be unable to guess what test solution is coming next.

Taste Sensation
Objectives:
- To demonstrate the gustatory sense.
- To understand there is more to the sensation of taste than just the individual sensations of sweet, sour, salty, and bitter.

Materials:
- Volunteers and experimenters.
- Paper towels.
- Paper cups.
- Cotton-tipped applicators.
- Disposable beakers, 5 ml (optional).
- Lab markers.
- Sweet solution (5% sucrose), salt solution (2% NaCl), sour solution (1% acetic acid), bitter solution (0.5% quinine), and distilled water.
- Oragel, Kank-a, or other oral, topical anesthetic agent (optional).

Procedure:

1. Select the subject and inform this person that he or she will need to stick the tongue out for this test. Also mention that you will be placing different solutions on different parts of the tongue and asking the subject to identify the solution as one of the four basic tastes.

2. Before testing, ask the subject to blot the tongue gently with a clean paper towel, and ask the subject to try to mentally identify each taste that will be tested before pulling the tongue back into the mouth in order to speak. Obtain small samples of each test solution in 5-ml, disposable beakers or paper cups. Mark the solutions, but do not allow the subject to see the markings.

QUESTION: Why is it necessary to identify the taste "mentally" before the tongue is pulled back into the mouth? (Hint: Consider what happens to the tongue when it is pulled back into the mouth.)

Observations:

Conclusion:

---

3. With a cotton-tipped applicator, moisten a small area on the tongue with one of the taste solutions. Allow a short period of time for the person to identify the taste. Out of view of the subject, record whether the response was correct. Do not record the response as being positive if given after the tongue is pulled back into the mouth.

4. Repeat the process until each test solution, and water as a control, have been tried on the tip, sides, middle, and rear of the tongue. Remember to mix the order of the applications and to use a different applicator for each taste. Allow the subject to drink, and possibly, rinse the mouth with plain water between tastes.

5. As an extension of this exercise, if time is available, test the effects of anesthesia on the sense of taste. Place a small drop of topical anesthetic designed for use in the mouth on one side of the tongue. After waiting until the tongue feels numb, test the ability of the person to detect tastes in the numbed area.

QUESTION: If you can still taste correctly even though your tongue is anesthetized, what does this tell you about the types of receptors the anesthetic actually affects?

Observations:

Conclusion:

Journal Note Think about the sensation of taste, or the gustatory sense. Think about all the wonderful taste sensations. There is a lot more to taste, however, than just sweet, sour, salty, and bitter. How do hot and cold affect tastes? How about fresh and stale? What do crisp and crunchy verses soft and mealy do to the taste of food? Are there any foods you do not like? If so, why do you not like them? Are there any foods that you have cravings for? How do you think food cravings relate to nutritional needs?

The Olfactory Sense

The sense of smell, or the olfactory sense, is located in the uppermost regions of the nasal passageways. The special area containing the receptors for smell is referred to as the olfactory epithelium. There are olfactory or hair cells, supporting cells, goblet cells, and submucosal mucous glands in this epithelium (Figure 8.3).

Smell sensations must be dissolved in the mucous layer for perception. For the gustatory sense, we have identified four distinct tastes, but there is much confusion regarding the number of primary smell sensations. Much of the evidence for different, discreet smell sensations comes from people who show "odor-blindness." This term describes individuals who cannot smell a particular odor, which implies this smell must be a discreet sensation that other individuals can smell.

Demonstrating the Olfactory Sense

There are few easy ways to demonstrate how the sense of smell functions. At best, students can be exposed to a variety of smells to determine if they can detect these particular odors.

Sense of Smell

Objectives:
- To demonstrate the sense of smell.
- To understand the strong mental images associated with the sense of smell.

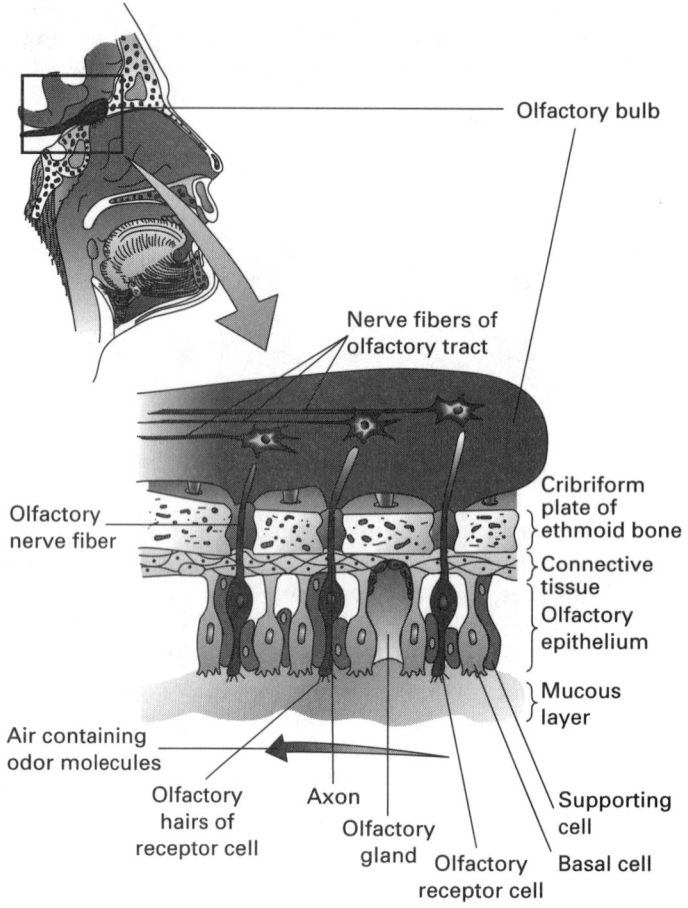

Figure 8.3
The Nasal Mucosa.
The components of the olfactory epithelium are hair cells, support cells, goblet cells, and mucous glands.

Materials:

- Odoriferous substances, such as pine-scented cleaner, lemon-scented cleaner or lemon juice, vanilla flavoring, peppermint-flavored mouthwash or peppermint flavoring, and rubbing alcohol.
- Petri dishes.
- Filter paper.
- Food coloring.
- Artificial flavorings, as available.

Procedure:

1. Prepare several petri dishes by placing a filter-paper wick in the smaller half of the dish. Be sure the wick does not extend above the edge of the lower half so that the dish can be covered.

2. Select a variety of chemicals or home solutions from the list of odiferous substances provided earlier, and ask the person to identify the odors. Pour the chemical/solution into the petri dish, and allow the wick to become completely moistened before testing. Unless there is any known or suspected risk associated with the chemical being tested, which would require the subject to control the position of the odor near his or her nose, this should also be a blind test. If subjects cannot specifically identify the compound, ask if they can categorize the smell as fruity, flowery, medicinal, chemical, unpleasant, and so on.

QUESTION: When entering a room, we often notice an odor that might be pleasant and appealing or repugnant and unpleasant. After a short time, we get used to the smell, and we are surprised to hear someone else entering the room remark about the odor. What is this "getting used to it" called, and why does this happen? (Hint: Consider what sensory value an odor might have. If the odor is not bad and not dangerous, is it worth being constantly aware of it? This is much the same as getting used to the sound of a dripping faucet or a rubbing fan blade.)

Observations:

Conclusion:

3. A variation of this test is to mix some special test solution and spike the odors with inappropriate coloration. The nose will perceive the odor, but if the visual cues are wrong, the subjsect may have difficulty with identification. Use food coloring and artificial flavors or extracts as the basis for this exercise, but do not use flavorings that are colored appropriately. (Almond, anise, and peppermint flavorings usually are colorless and, therefore, could be colored inappropriately.)

4. Set up petri dishes as described earlier with filter-paper wicks, and test whatever flavoring and colors are available.

QUESTION: We have definite mental images regarding what certain foods, flavorings, and so on should look and smell like. What is stronger, the visual cues (i.e., the colors) or the odor cues (i.e., the smells)? (Hint: Remember the mental images based on fond childhood memories.)

Observations:

Conclusion:

Journal Note How good is your sense of smell? Can you identify odors correctly? Do women have a better sense of smell than men? Do any odors make you sick? Did you know that natural gas is odorless, but that utility companies add a chemical to the gas supply so that we can smell leaks? What is the chemical added to the gas supply?

The Visual Sense

The eyes are very specialized structures that are designed to detect changes in the level of light. All would agree that the eyes allow us to do many other valuable things.

Structure of the Eyes

Each eye is located in a depression on the anterior side of the face, which is called the **orbit** (Figure 8.4). The eye or eyeball consists of three layers. One is an external **sclera** (*sclero* (G), hard), which is a layer of connective tissue that gives the white appearance to the outside of the eyeball. On the anterior of the eye, the sclera becomes transparent; this is called the **cornea**. The second layer is the **choroid** (*choroid* (G), like a membrane). It is rich in blood vessels and pigment and is an important source of blood-borne nutrients for the wall of the eyeball. In the anterior portion of the eyeball, the choroid gives rise to the lens and its associated structures, such as the **suspensory ligaments,** and the **ciliary body. The** third and inner layer of the eye is the **retina** (*retina* (L), a net), which consists of the sense cells that detect the light that is perceived by the eyes (Figure 8.5). The sense cells are the **rods** for low-light vision and **cones** for color vision. There actually are several layers that make up the retina as well. From the choroid layer toward the center of the eyeball there is the **pigment layer** (cells), the rods and cones with the sensory part facing the pigment cells, the **bipolar cells,** and the **ganglion cells.**

The eye's interior is divided into an **anterior cavity,** which is the area anterior to the lens, and the **posterior cavity,** which is the area from the posterior surface of the lens to the retina. The axons of the ganglion cells merge at a point on the posterior of the eye to form the **optic nerve,** which is a sensory nerve to the brain. The area where the optic nerve exits the eyeball is called the **blind spot,** or **optic disc,** because there

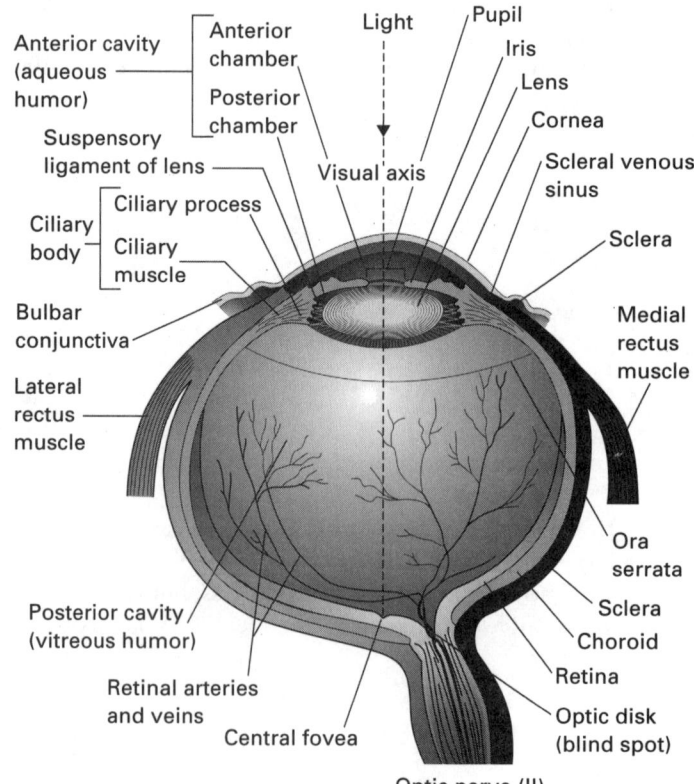

Figure 8.4
The Eye.
The components of the eye are shown in this sagittal section of the eye. Note the following structures: orbit, sclera, cornea, choroid, suspensory ligaments of the lens, ciliary body, retina, rods, cones, pigment layer, bipolar cells, ganglion cells, anterior cavity, anterior chamber, posterior chamber, posterior cavity, optic nerve, optic disc, macula lutea, and fovea centralis.

Figure 8.5
The Retina.
The layers of the retina are shown in these cross-sections. Note that the photoreceptors, the rods and cones, face away from the source of light. They are connected to the bipolar cells, which are connected to the ganglion cells, whose axons actually form the optic nerve.
(a) Schematic representation.
(b) Photomicrograph.

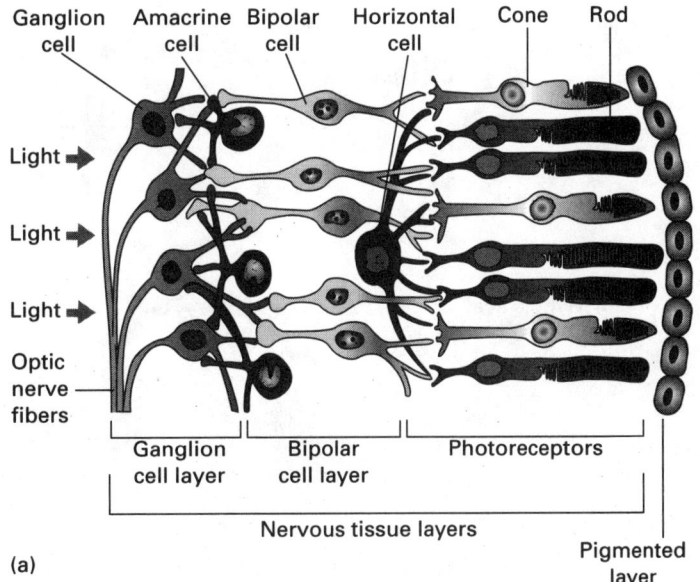

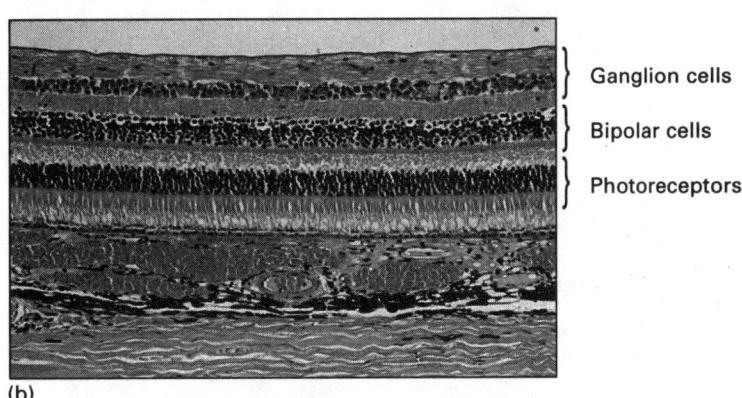

are no rods and cones in this area. Near this area also is a smaller area that contains nothing but cones and, therefore, represents a spot of best vision; this is called the **macula lutea** (*macula* (L), spot; *lute* (L), yellowish). This spot has fewer blood vessels covering it, and the bipolar cells and ganglion cells also are displaced so that the area appears to be lighter-colored compared to the blood vessel–rich areas surrounding it. This area, and especially the **fovea centralis** (*fovea* (L), a pit), which is a depressed spot in the center of the macula, is the point of best focus.

Determination of Visual Acuity

Visual acuity refers to the sharpness of the image that we see. The ability to adjust the shape of our lens so that we can focus on objects at different distances is called **accommodation**. **Emmetropia** refers to normal vision, in which the image is focused on the retina (Figure 8.6). If the image is focused in front of the retina and distant objects are out of focus, the condition is called **myopia**, or **nearsightedness**. When the image is focused behind the retina, near objects are out of focus, and the person has **hyperopia**, or **farsightedness**.

Figure 8.6
Visual Acuity.
This illustration shows where near and far objects are focused on the retina. In normal vision, or emmetropia, all objects are focused on the retina. In myopia and hyperopia, images are focused either in front of or behind the retina, respectively, and, therefore, are not in focus.

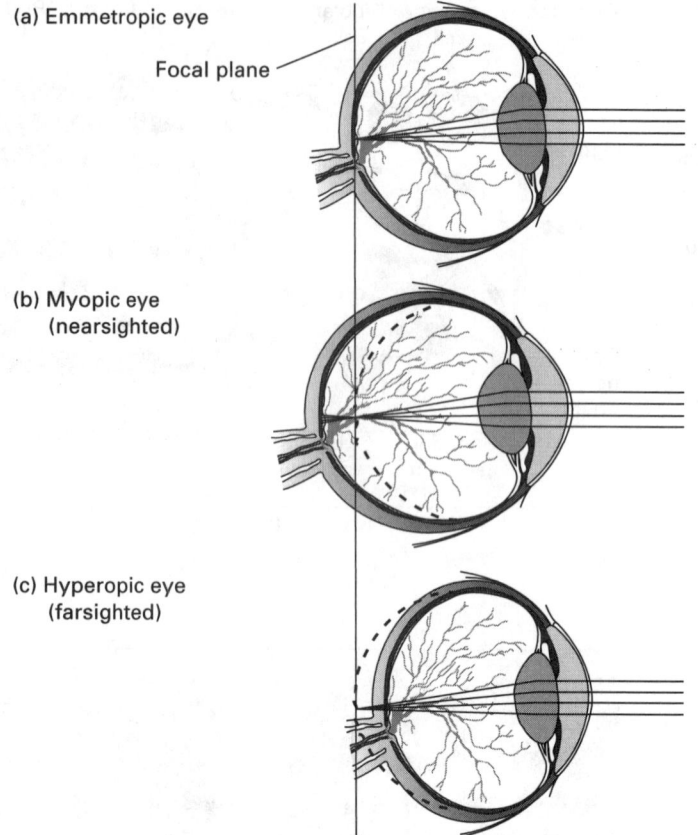

(a) Emmetropic eye
   Focal plane

(b) Myopic eye (nearsighted)

(c) Hyperopic eye (farsighted)

Determining Visual Acuity

Objectives:

- To understand the function of the Snellen eye chart as a device to determine visual acuity.
- To determine visual acuity.

Materials:

- Snellen eye chart or comparable acuity chart.
- Twenty-foot tape measure.
- Pointer (optional).
- Opaque card (optional).

Procedure:

1. Place a Snellen eye chart on the wall in a spot that allows the subject to stand 20 feet away (Figure 8.7). A distance of 20 feet guarantees the light rays are parallel when they strike the cornea of the eye.

2. Cover the eye not being tested with a hand or an opaque card (Figure 8.8). (Do not simply close the eye not being tested by muscular action of the eyelid.) Have the experimenter point to a line on the chart and ask the subject to read that line. If the line is read correctly, ask the subject to read the next-smallest line. Continue this procedure until the subject makes errors or cannot read the line at all. Record this reading for that eye.

3. Cover the other eye, repeat the procedure, and record the acuity. Now try the test with both eyes. Are the eyes similar in acuity? If the subject wears glasses, try the test both with and without the glasses.

Figure 8.7
The Snellen Chart.
The Snellen chart is the standard eye test chart. When standing 20 feet from the chart and with one eye covered, an individual's ability to accurately read the chart can be determined. A person who can read what an average, normal individual can at 20 feet is said to have 20/20 vision.

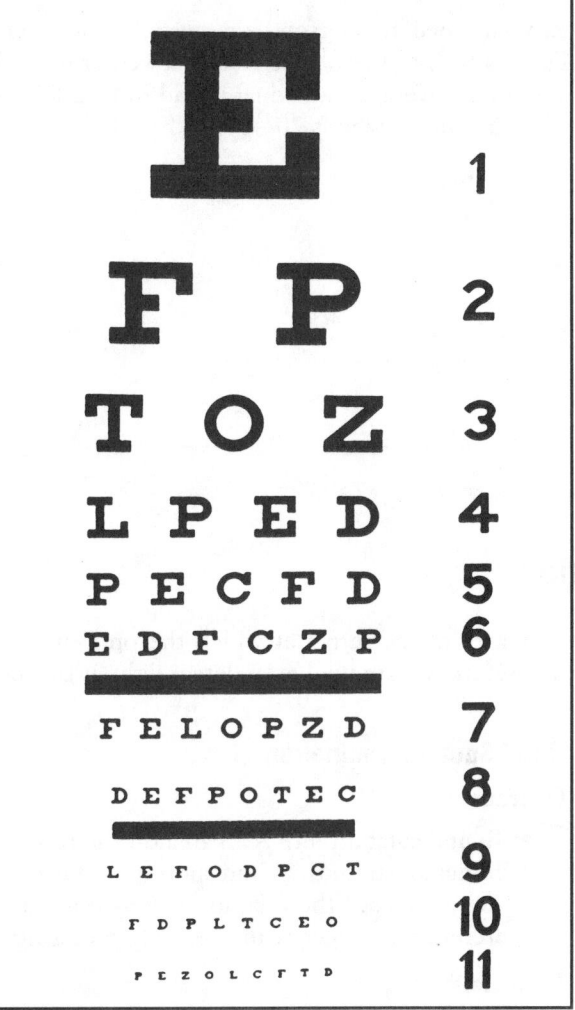

Figure 8.8
Checking Visual Acuity.
This person is shown covering one eye to read an eye chart. Eyes should be covered, not "squinted shut," because this distorts the face and may affect the results.

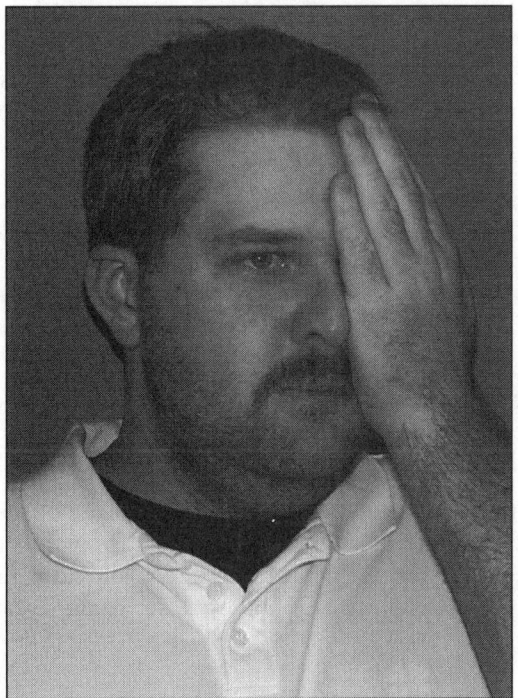

Determination of Visual Acuity 127

QUESTION: As mentioned, the person being tested stands 20 feet from the chart to guarantee that light rays from the object being viewed are parallel. Thus, someone who can see as well as the average individual is said to be 20/20. How would you describe the vision of an individual who is 20/400? What about an individual who is 20/15?

Observations:

Conclusion:

Blind Spot Determination

The area of the retina at which the optic nerve exits the eyeball has no receptors. Therefore, we are unable to detect light at this point. This is the blind spot.

Blind Spot Determination

Objectives:
- To understand there is an area on the retina that contains no light receptors.
- To demonstrate the blind spot in each eye.
- To understand there is no obvious blind area in the visual field because of the presence of two eyes and central processing.

Materials:
- Opaque card (optional).
- Blind spot test card, or the pattern that follows.

Procedure:

1. To demonstrate the blind spot, use the pattern displayed here or, draw a similar pattern on a blank index card:

    ●   +

2. Cover one eye, and test the uncovered eye. Hold the card or book so that the cross is directly in front of the eye being tested, with the dot positioned laterally. Start with the card at arm's length, then move the card closer to the eye.

3. Keep the head erect and the card in the plane of the eye. Stare at the cross continuously, but note that the dot disappears from your awareness. Keep looking at the cross. While not staring at the dot, you should realize that it goes out of view and then reappears as the card comes closer to the eye.

4. As you move the card closer or further away, light is reflected from the surface of the card to the eye. As the card is moved, the light reflected from the area with the dot passes across the retina. When this light impinges on the retina, where there are no receptors, the dot is not seen.

QUESTION: If the blind spot is a circular area on the retina, that means there is a circular area in your visual field in which you cannot see anything. Why do you not notice this big empty space (the blind spot) on each side of you?

Observations:

Conclusion:

Journal Note Do you like 3D movies? How are 3D effects generated? To help you figure this out, consider that most people have two eyes, and that the eyes are located at a slightly different location on the face. They are seeing essentially the same image, but it is slightly displaced. This is why we have binocular vision. How do the special 3D glasses allow us to see 3D images?

Color Blindness

Based on the protein associated with the molecule retinene, which is the same pigment found in the rods, there are thought to be three types of cones. Because of these different proteins, the three different cones are sensitive to red, green, or blue wavelengths of light. A genetic lack of one or more types of proteins results in the inability to perceive light of that color, which is called **color blindness.** The most common type is red-green color blindness.

Color Blindness Test

Objectives:
- To learn how simple color-blind testing is performed.
- To ascertain if the subject has any color-detection deficiency.

Materials:
- Color-perception test cards.

Procedure:

1. Obtain color-blindness test charts from your instructor.

2. Look at the test chart, and read the letter or number in the field of dots.

3. Refer to the available charts to determine if the subject is color blind.

QUESTION: There are different types of color blindness, with the most common being the inability to see red and green. Consider all the devices in everyday life that use the colors red and green. How can color-blind individuals deal with this disorder?

Observations:

Conclusion:

The Auditory Sense

The sense of hearing is associated with the inner ear, which also is involved with the sense of balance and equilibrium. The ear is divided into the **external ear,** the **middle ear,** and the **inner ear** (Figure 8.9).

The external ear consists of the **pinna** (*pinna* (L), a wing) or **auricle** (*auricul* (L), the ear), which is the flap on the side of the head that wraps around the opening of the **external auditory canal.** The pinna catches the sound waves, and it directs them down the ear canal. At the end of the external auditory canal is the **tympanic membrane.** This membrane is a thin layer of connective tissue, which is covered with skin on the outside and mucous membrane on the inside, that vibrates when sound waves hit it. The tympanic membrane is the partitioning structure between the external and middle ear.

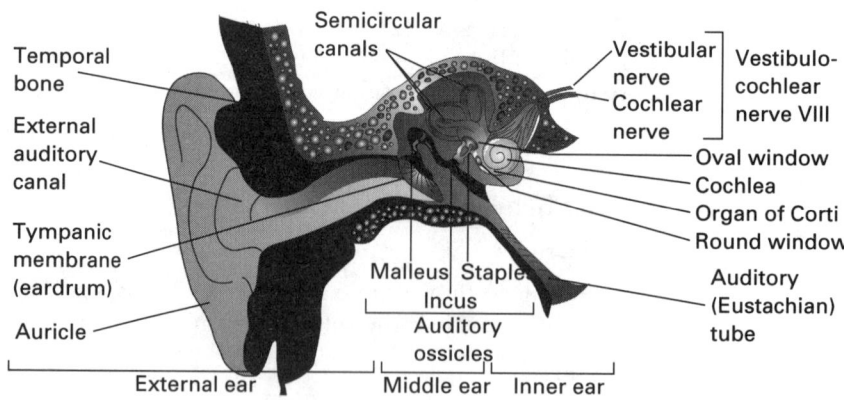

Figure 8.9
The Ear.
This cross-section of the ear shows the external, middle, and inner sections. Note the following structures: external ear, middle ear, inner ear, pinna, meatus, tympanic membrane, malleus, incus, stapes, oval window, membranous labyrinth (including utricle, saccule, semicircular canals, cochlea), bony labyrinth, ampulla, organ of Corti, and cochlea.

130 SECTION 8 The Nervous System—Part 2

QUESTION: The pinna catches sound waves and funnels or directs them down the external auditory canal and toward the tympanic membrane. How do people increase their ability to hear by modifying the shape or dimensions of the pinna?

Observations:

Conclusion:

The vibrating membrane causes the **auditory ossicles, malleus** (*malleus* (L), a hammer), **incus** (*incus* (L), an anvil), and **stapes** (*stapes* (L), a stirrup) to vibrate in sequence, thus passing the sound information through the middle to the inner ear. The opening into the inner ear to which the stapes is attached is called the **oval window,** and this allows the vibrations in the middle ear to pass to the inner ear. The inner ear holds a fleshy structure that contains the organs of audition and equilibrium. This complicated part of the inner ear is referred to as the **membranous labyrinth** (*labyrinth* (G), a maze). Four distinct structures make up the membranous labyrinth, which are the **utricle** (*utric* (L), a leather bag), the **saccule** (*sacc* (L), a sack), the **semicircular canals,** and the **cochlea** (*cochlea* (L), a snail shell). (The utricle and saccule are collectively called the **vestibule.**) The bone that surrounds these structures has about the same shape as the membranous labyrinth and is called the **bony labyrinth.** The utricle and saccule contain the receptor for static equilibrium and linear acceleration, or the **macula** (*macula* (L), spot). The semicircular canals contain the receptor for dynamic equilibrium or angular acceleration, or the **cristae** (*crist* (L), a crest), in an enlarged area at the base of each semicircular canal, or the **ampulla** (*ampulla* (L), a flask). The cochlea contains the receptor for hearing, or the **organ of Corti.**

Figure 8.10
The Organ of Corti.
This cross-section of the organ of Corti shows the arrangement of the hair cells that are receptors for the sense of hearing. Note the following structures: scala vestibuli or vestibular duct, perilymph, scala tympani or tympanic duct, round window, cochlear duct, endolymph, vestibular membrane, tectorial membrane, and basilar membrane.

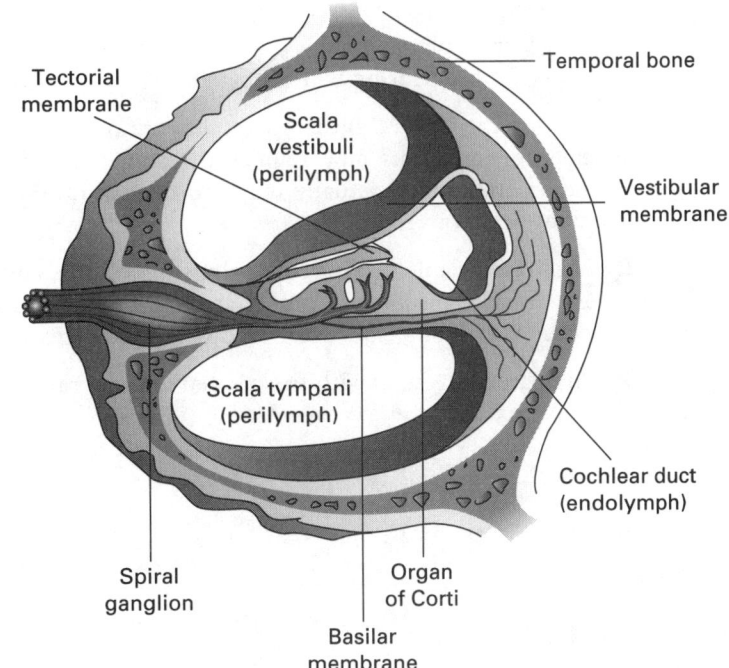

The structure of the cochlea and organ of Corti require further elaboration, because they are quite complicated (Figure 8.10). The cochlea actually is divided into three channels. A **scala vestibuli** (*scala* (L), a ladder), which contains **perilymph,** connects with the oval window and extends to the tip or end of the cochlea, where it connects to the **scala tympani,** which, in turn extends back to the beginning of the cochlea, where it connects with the **round window.** Separating these two channels is the **cochlear duct,** which has the **vestibular membrane** as its "roof" and the **basilar membrane** as its "floor." The organ of Corti "sits" on the basilar membrane, is covered by the **tectorial membrane,** and is bathed by **endolymph.** As sound waves are transmitted from the tympanic membrane, the fluid in the scala vestibuli conducts the vibrations to the part of the basilar membrane that is responsive to that wavelength. The organ of Corti on the basilar membrane moves and rubs the hairs of the organ against the tectorial membrane and, thus, sends signals to the brain.

The Hearing Test

A device called an **audiometer** is used to test hearing ability. The subject dons a headset and, without looking at the controls, is asked to indicate when and in which ear a sound is heard. The experimenter varies both the frequency and the volume or amplitude of the sound produced by the audiometer. The lowest volume perceived at each frequency is recorded and then compared with normal values.

The Hearing Test

Objectives:

- To understand the sense of hearing.
- To appreciate the techniques involved in accurately measuring the ability to hear.

Materials:

- Volunteers.
- Audiometer.
- Audiometer score sheets (optional).

Procedure:

1. Find a quiet place, such as a lab prep room or separate lab/classroom. Select the subject whose hearing will be tested.

2. Position the subject so that he or she cannot see the controls or movements of the experimenter. This may require the subject to sit with his or her back to the audiometer, which actually may prevent confusion on the experimenter's part in telling left from right.

3. With the variable or discrete settings for volume and frequency, establish the lowest volume the subject can hear at each frequency tested. There may be a printed score or recording sheet provided with the particular audiometer being used.

4. Compare the results with the normal hearing range for humans.

QUESTION: Like many other capabilities, hearing changes with age. Some abilities, for example, improve and then decline. What changes occur in our hearing as we age? Are there any older individuals in the class who show a lessened ability to hear?

Observations:

Conclusion:

Journal Note There are many pieces of equipment in our environment these days, and they all make some noise. Even some light bulbs make noise. How noisy is your environment? This "background" noise sometimes is called "white" noise. How can we make our surroundings less distracting? Do you like noise in the background?

Hearing Deficit and Diagnosis

There can be many reasons for hearing deficiency, but there generally are two types of hearing loss. Sensorineural loss or deafness results from damage by birth, injury, or disease to the organ of Corti or vestibulocochlear nerve. Conduction loss or deafness results from damaged or defective middle ear ossicles. A couple of simple tests differentiate these two types of deafness. The Weber's test is performed with a tuning fork placed in the middle of the forehead and records whether the sound is equal or louder in one or the other ear. The Rinne test involves placing the end of the tuning fork on the mastoid process and then in front of the external auditory canal; whether the sound is heard loudest at the canal or through the bone differentiates sensorineural from conduction deafness. Tuning forks are Y-shaped devices that vibrate to produce a particular frequency sound when struck. They can be made in any frequency for testing purposes, for tuning musical instruments, or for calibrating electronic equipment.

Weber's Test

Objectives:
- To understand the procedure used in the Weber's test.
- To appreciate the Weber's test as a means to determine hearing loss.

Materials:
- Tuning fork.
- Rubber mallet or hammer.

Procedure:
1. Select the subject to be tested, and ask that he or she to close the eyes. In as quiet a setting as possible, gently strike the tuning fork with a rubber mallet and hold it in front of the face, touching the end to the middle of the forehead (Figure 8.11).

Figure 8.11
Weber's Test.
Note the positioning of the tuning fork in the center of the forehead.

2. Ask the subject to indicate if the sound is louder in one ear or the same in both.

3. The results are stated as "Weber negative" if the sound is the same in both ears. "Weber right" and "Weber left" indicate if it is louder in the right or left ear, respectively.

Weber negative   Yes   No (if no, indicate if Weber right or Weber left.)

Right ear _____   Left ear _____

If the sound is louder in one ear, that indicates there is some hearing deficit.

4. Repeat the test, this time placing a finger in one ear and then the other. If the ear that is blocked perceives the sound better than the unblocked one, then the blocked ear is normal. If the unblocked ear perceives the sound better than the blocked ear, there may be hearing loss in the blocked ear.

Rinne Test

Objectives:
- To learn to perform the Rinne Test.
- To appreciate the use of the Rinne Test in a basic auditory workup to determine the type of hearing loss.

Materials:
- Tuning fork.
- Rubber mallet or hammer.
- Stopwatch.

Procedure:

1. Mask one ear, and gently strike the tuning fork with a rubber mallet. First, place the stem of the tuning fork on the mastoid process behind the other ear, then place it a half-inch from the external auditory canal (Figure 8.12).

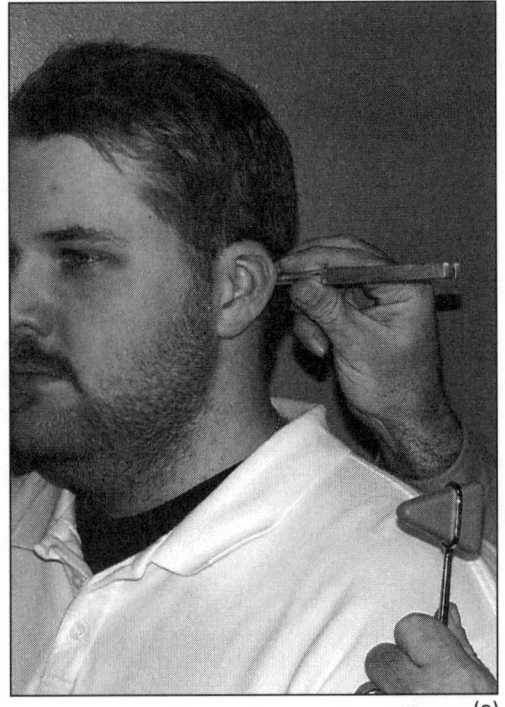 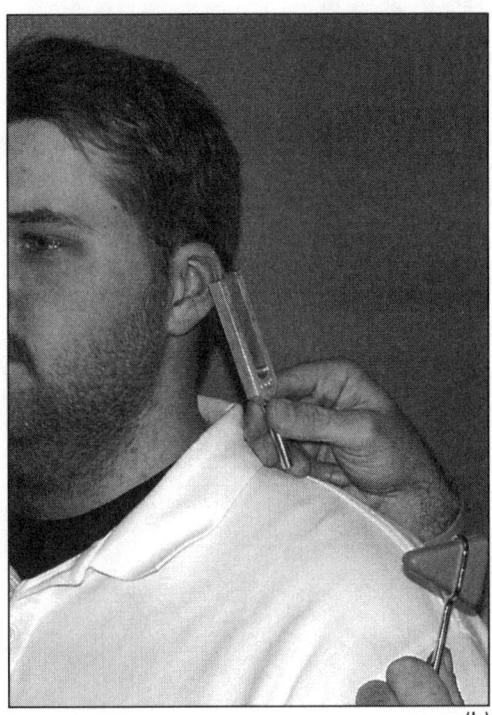

(a)   (b)

Figure 8.12
Rinne Test.
(a) Note the position of the tuning-fork end on the mastoid process behind the ear.
(b) The tuning fork then is moved to a position at the opening of the external auditory canal.

2. After the sound has died away at either position, move the tuning fork to the other position, and note if the sound is heard.

3. If the sound is heard longer at the auditory canal than at the bone, this is referred to as "Rinne positive." This condition indicates normal hearing or possibly sensorineural deafness. If the sound is heard equally by bone or air conduction, this is referred to as "Rinne equal." If the sound is heard longer via bone than through air conduction, this indicates conduction deafness and is called "Rinne negative."

	Right ear	Left ear
Rinne positive	_____	_____
Rinne equal	_____	_____
Rinne negative	_____	_____

QUESTION: In either of the two tests discussed, will the frequency of the tuning fork being used have any effect on the outcome?

Observations:

Conclusion:

Hearing Deficit and Diagnosis   **135**

Section 9

The Circulatory System

Note to the Student

The circulatory system is composed of several recognizable entities. The heart, blood vessels, and blood are what come to mind whenever the circulatory system is mentioned. The system is a closed set of tubes, which contain a certain volume of fluid that is moved throughout the system by a muscular pump. The muscular squeeze of the heart and the resistance from the tone or slight contraction of the smooth muscle in the vessel walls maintain blood pressure. Without the pumping action of the heart or the resistance provided by the vasculature, there is no flow and no pressure. The most common form of shock or very low blood pressure occurs when contraction of the smooth muscle is interfered with, resulting in a loss of peripheral resistance and fall in blood pressure. The person feels light headed and may even faint, because the brain is not receiving an adequate amount of blood. The major factor in control or maintenance of blood pressure is to guarantee an adequate blood supply to the brain.

Architecture of the Circulatory System

The circulatory system is composed of two circuits or loops (Figure 9.1). The pulmonary circuit carries blood that has just returned from the body to the lungs, where the blood picks up oxygen and gives up carbon dioxide. The blood then returns to the heart and is pumped to the entire body to supply the tissues with necessary nutrients and oxygen; this is called the systemic circuit. There is only one pump, the heart, but it has two sets of chambers. Therefore, the two circuits remain separate. Each side of the heart contains both a receiving and a pumping chamber (Figure 9.2). The receiving chamber, or the atrium (*atrium* (L), vestibule; plural, atria), is thin walled and does not contain much muscle. Many students assume that because part of the heart is very muscular, then all of the heart is muscular. The larger chamber, or the pumping chamber, on each side of the heart is called the ventricle (*ventricul* (L), belly). The walls of the left ventricle are much more muscular than those of the right. Both circuits use the same scheme of distribution, however, with large, thick-walled vessels receiving blood as it is pumped by the heart and that taper gradually as the blood travels further from the heart. They also branch extensively as they travel away from the heart, to supply all tissues with what they need. Then, the blood is returned to the heart.

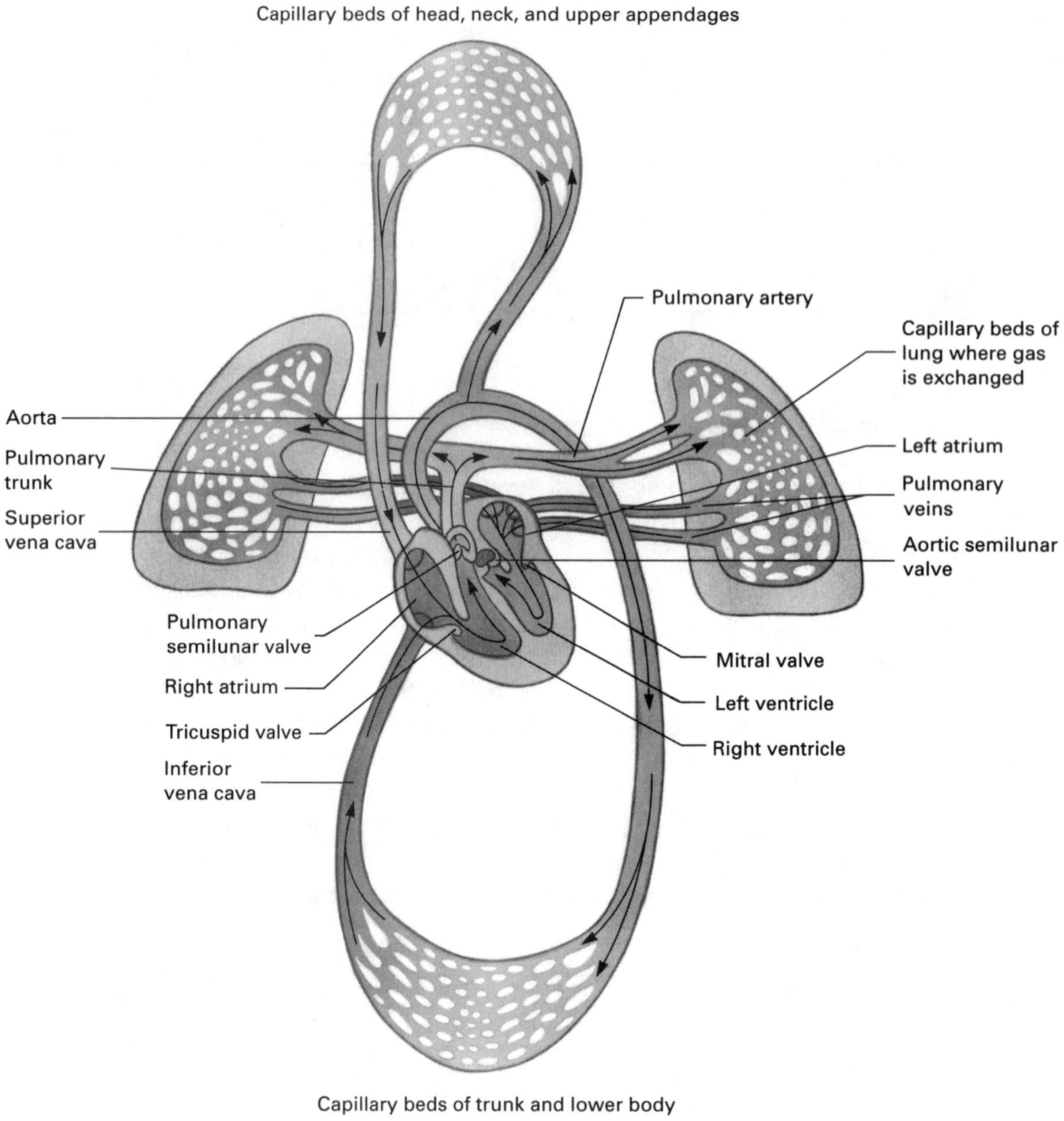

Figure 9.1
The Heart, Pulmonary Circuit, and Systemic Circuit.
These are the principal components of the circulatory system: a pump, the heart, and two sets of blood vessels. One set of blood vessels oxygenates the blood by passage through the lungs, and one set distributes the blood to the rest of the body.

Figure 9.2
The Heart and Large Vessels.
This illustration shows the anterior surface of the heart and the large vessels carrying blood to or away from the heart.

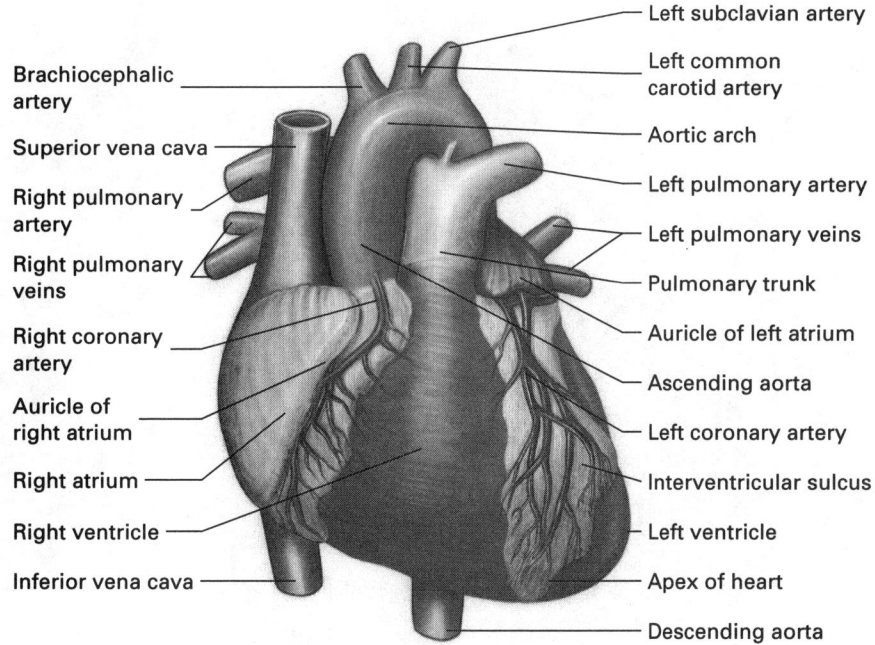

All vessels that carry blood are described as having three layers in the vessel wall (Figure 9.3). The innermost layer is called the **tunica intima** or **interna** (*tunic* (L), a covering; *intima* (L), innermost). This layer consists of an inner **endothelium,** its basement membrane, and a connective tissue layer that may contain quite a bit of elastic fibers. The name endothelium refers to the simple squamous epithelium that lines the blood vessels and the heart. The middle layer of the vessel is the **tunica media** (*medi* (L), the middle), and it contains the smooth muscle of the vessel as well as connective tissue. This layer that maintains the tone of the system to regulate blood pressure. The outermost layer is the **tunica adventitia** or **externa** (*extern* (L), outer, outside); it is composed mostly of connective tissue and reinforces the vessel wall.

The large vessels that carry blood away from the heart are called the **arteries** (*arteri* (G), an artery). Individual arteries are named either for the areas of the body or organ they supply. Arteries in general are referred to as conducting vessels, because they bring blood from the heart to the tissues. The vessels branch and get smaller until they are essentially an endothelium, its basement membrane, and a layer of smooth muscle cells of the media. These vessels are the **arterioles** (*-iol* (L), little), and they give off branches that are the smallest vessels in the system. These vessals are called the **capillaries** (*capill* (L), a hair). They are composed of an endothelium and a basement membrane. This simple arrangement is the thinnest that a biological structure can be to facilitate functioning of the capillaries. They pass valuable materials in the blood to the tissues and waste products from the tissues to the blood; thus, capillaries are called exchange vessels.

As blood flows out of the capillaries, it begins on its journey back to the heart in the **venules** (*ven* (L), a vein; *-ule* (L), little). The venules join together and increase in size as they drain more blood from the tissues and carry it back to the heart. The larger vessels are called **veins.** We start to the see the same three layers associated with blood vessels in the venules and veins. The veins have less elastic connective tissue and less smooth muscle than comparable arteries and arterioles. The veins also tend to hold or allow blood to pool; hence, they are called capacitance vessels. Finally, all the veins make their way toward the heart, and they deliver the blood to be pumped again on another circuit through the body.

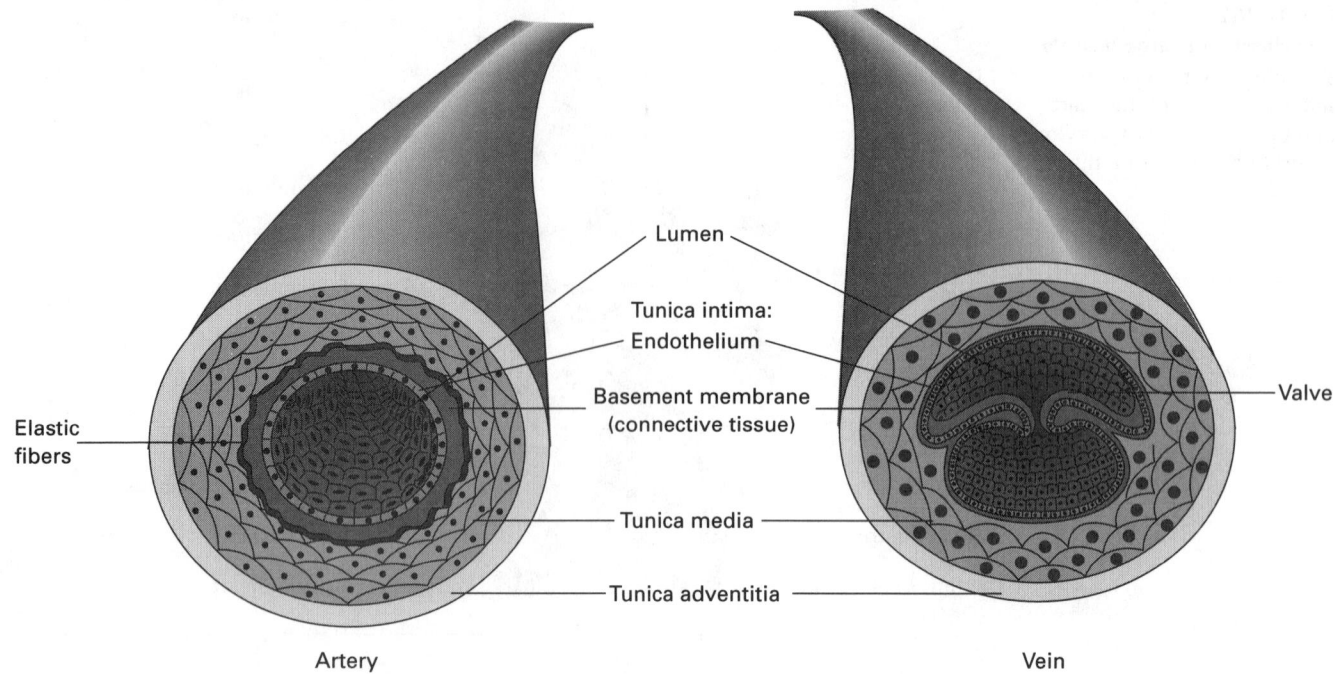

Figure 9.3
An Artery and Vein.
Note the difference between the thick-walled artery with a small lumen and the thin-walled vein with a large lumen.

When first pumped, blood is under pressure, and that pressure is maintained by the integrity of the circulatory system. After passage through the capillaries, blood is under very little pressure. Therefore, it is difficult to get the blood to return to the heart. This is especially a problem with return of blood from the feet, because this return is against gravity. To assist in the return of blood to the heart, there are three structures or mechanisms: valves in the veins, the muscular pump, and the partial vacuum in the thoracic cavity with breathing. First, there are one-way valves in the veins so that once blood has been squeezed past a valve, it does not go backward (Figure 9.4). The term muscular pump does not refer to the smooth muscle in the veins but to the skeletal muscles that surround and envelope the deep veins in the extremities. When these muscles contract as they move parts of the body or maintain posture, they shorten, and the belly of the muscle gets fatter and pushes on or squeezes the surrounding structures (Figure 9.5). This includes the veins and so the blood is pushed back to the heart. Finally, during breathing, movement of the thoracic muscles (especially the diaphragm), creates a partial vacuum in the thoracic cavity, which pulls air in for inspiration. This drop in pressure also helps blood get back to the heart.

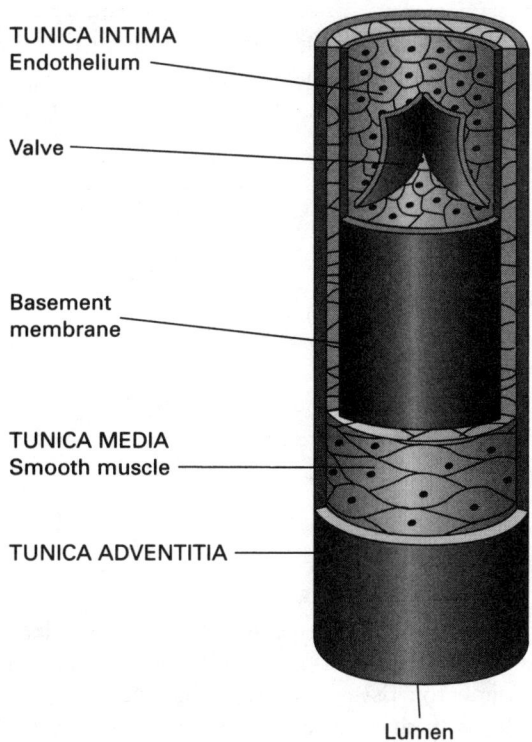

Figure 9.4
A Valve within a Vein.
One-way valves in the vein are one mechanism that assist in return of blood to the heart.

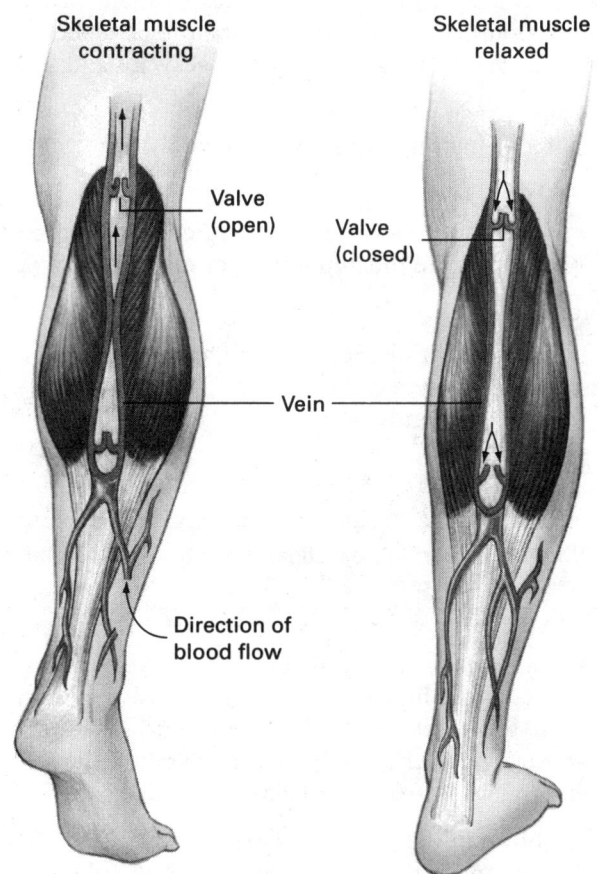

Figure 9.5
The Muscular Pump.
The deep veins of the body are surrounded by skeletal muscles. When these muscles contract, they squeeze blood past the valves in the veins.

Architecture of the Circulatory System

QUESTION: Why are band members and military recruits warned to "not lock their knees" when standing at attention for any length of time?

Observations:

Conclusion:

Demonstrating the Valves in the Veins

There are two sets of larger veins in the arms and legs: the deep veins, and the superficial veins. The deep veins are inside the arms and legs, and are surrounded by muscles and connective tissue. The superficial veins are evident on the surface of the body, beneath the skin; they are especially obvious on the forearms and lower legs in very slim or lean individuals. Unless an individual is obese or very swollen, superficial veins can be found on the tops of the feet and the back of the hands. The following exercise is an adaptation of the same experiment performed by William Harvey. This information helped him to understand that blood flowed in a circular path through the blood vessels rather than just a backward and forward ebb and flow.

Demonstrating the Valves in the Veins

Objectives:
- To demonstrate one of the differences between arteries and veins: the valves.
- To appreciate the importance of valves in conveyance of blood back to the heart.

Materials:
- A subject with large or obvious superficial veins.
- An experimenter.
- Antibiotic handwash.

Procedure:

Note: It is good lab practice, though not essential, to wash your hands before starting this exercise.

1. Examine the inside of the forearms, lower legs, or tops of the feet of individuals in the class, and find a student with obvious superficial veins. Restrict the blood flow from these vessels with a loose tourniquet or the opposite hand around the appendage. Clench and open the hand on the selected arm or move the foot several times to expand the veins with blood.

2. As shown in the photo (Figures 9.6a through 9.6d), use the thumb to block a vein distally, and with the first finger, push the blood past a valve. The vein will remain collapsed until you lift the thumb.

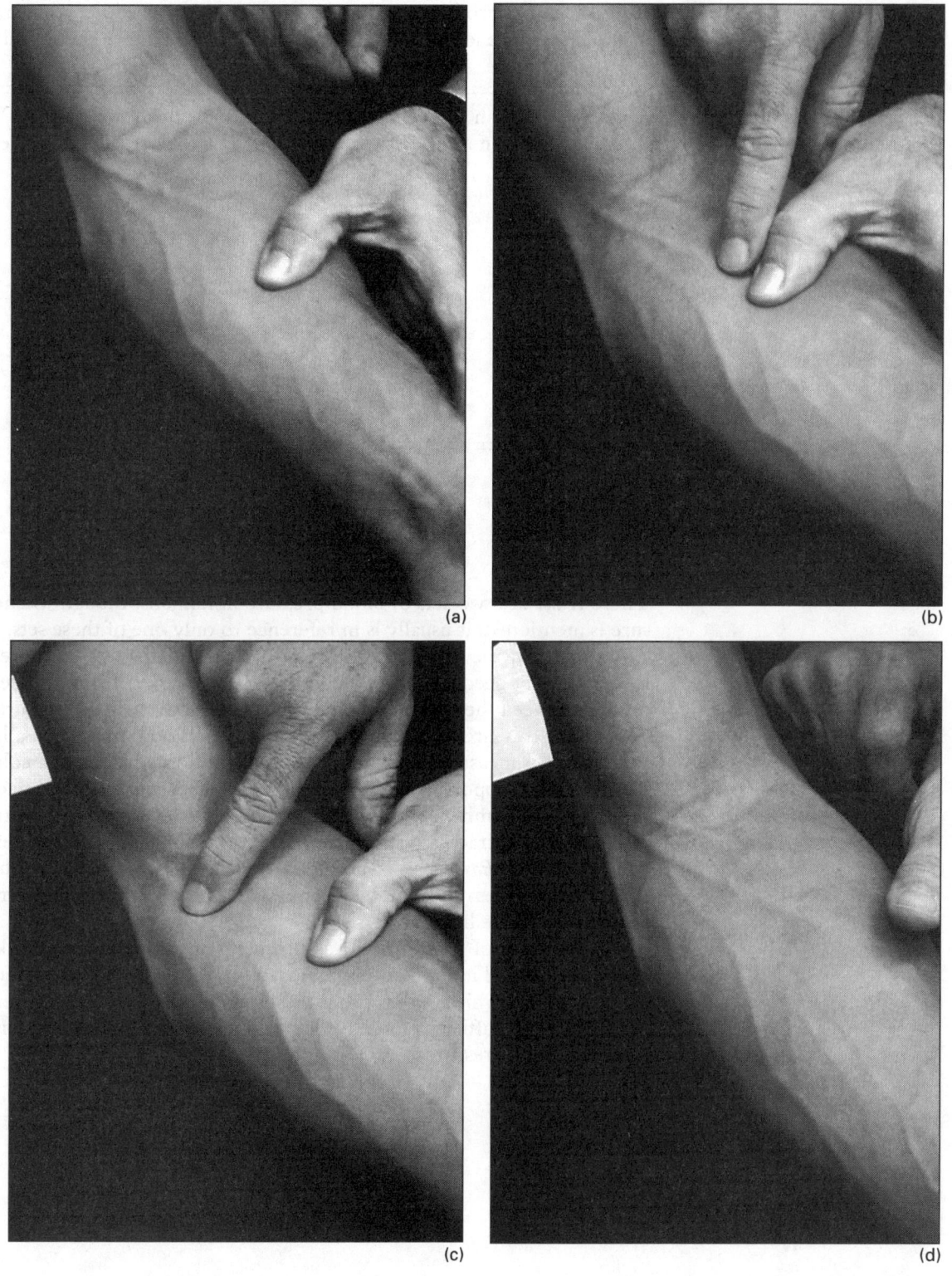

Figure 9.6
Demonstrating Valves in the Veins.
(a) One hand can be used to partially restrict the flow of blood out of the arms via the superficial veins. (b) The thumb of the other hand can be used to block the flow of blood in the selected superficial vein. (c) Another finger is used to push the blood out of the vein past a valve. (d) The vein should be collapsed between the thumb and the valve. (Note: the vein should remain collapsed until the thumb is removed. If the site selected does not work, try another location.)

3. If you do not produce a collapsed vein with this procedure, you may not be pushing the blood far enough past a valve. Reposition the finger and thumb and try again.

QUESTION: Valves are found in the veins and the heart, but not in the arteries. What is the key difference between arteries and veins relating to the presence of valves in the veins?

Observations:

Conclusion:

Measuring Blood Pressure

There really are two sets of blood pressure in humans, though whenever blood pressure is mentioned, it usually is in reference to only one of these sets. Just as there are two circuits in the circulatory system, each circuit has its own pressure. The pulmonary circuit needs to pump blood from the heart to the lungs. Because the heart is nestled between the lungs, however, this takes relatively little pressure, thus the blood pressure in this circuit is low ($\approx$ 25/8 mm Hg). Blood pressure is a hydrostatic pressure, and it is measured as other pressures are: as the height of a column of mercury the blood can support based on pumping of the heart and squeezing of the blood vessels. The first number (25) is the **systolic pressure** (*systol* (G), a contraction); **systole** refers to the contractile phase of the heart cycle. The second number (8) is the **diastolic pressure** (*diastol* (G), standing apart). The diastolic pressure, then, refers to the pressure the system drops to between heartbeats, with **diastole** referring to the relaxation phase of the heart cycle.

A pressures of 25/8 mm Hg is nowhere close to what we think of as a normal blood pressure. Thus, when discussing blood pressure, we are referring to the systemic pressure, which is that produced by the more muscular left ventricle to propel the blood with sufficient force to deliver a constant supply to the brain. The typical systemic blood pressure is 120/80 mm Hg.

QUESTION: Most blood pressure measurements are determined with a cuff device, which is placed around the upper arm. This is how systemic blood pressure is measured. Could this same method be used to measure pulmonary blood pressure?

Observations:

Conclusion:

Determining Blood Pressure

Blood pressure can be measured either directly or indirectly. Direct blood pressure measurements are done only in surgical settings or animal experimentation. It typically involves insertion of some catheter-like pressure sensor, which is connected to a recording apparatus that can amplify and record the signal. This is an invasive procedure and is much too dangerous for routine monitoring; therefore, almost all blood pressure determinations are indirect. Indirect measurements use an auscultation method that involves listening to sounds arising from the movement of blood inside the body. The stethoscope is a listening device that helps us to hear these sounds. Blood flow in an unrestricted vessel is silent, however, because it is nonturbulent. For blood flow to be detectable, an inflatable cuff with a pressure gauge attached is used to restrict the lumen of the vessel and cause noisy, turbulent flow. The cuff device is called a sphygmomanometer (Figure 9.7). As the pressure in the cuff is decreased, some blood can be forced through the previously occluded vessels, thus causing a noise that can be heard.

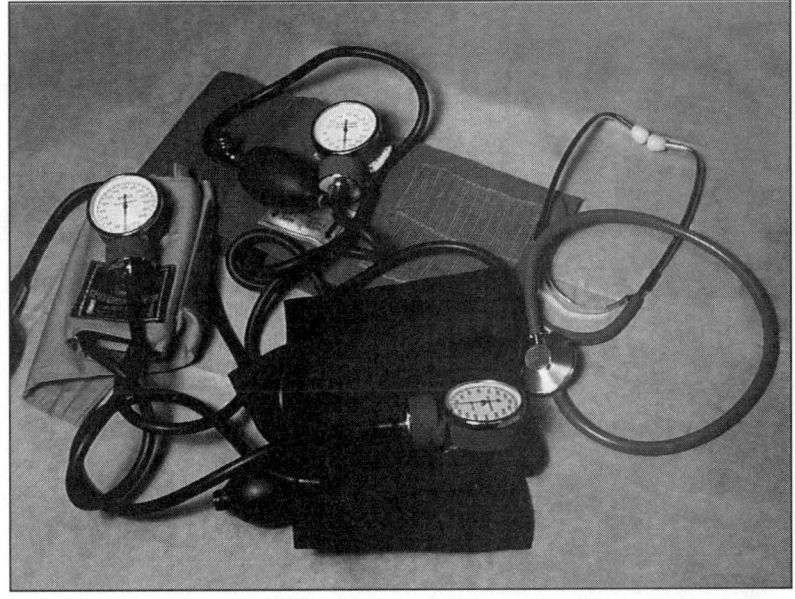

Figure 9.7
Different Types of Sphygmomanometers.
These are various types of sphygmomanometers, or blood pressure cuffs, that may be used.

Determining Blood Pressure

Objectives:

- To understand the process for determining systemic blood pressure with the auscultation method.
- To realize that nonturbulent blood flow is silent.
- To learn to take a blood pressure reading with a sphygmomanometer and stethoscope.

Materials:

- Sphygmomanometer.
- Stethoscope.
- Alcohol pads.

Caution: It is advisable to clean the earpieces of the stethoscope with an alcohol pad before use.

Procedure:

1. Select the arm to be used for the blood pressure determination. Expose the inside of the elbow on this arm.

2. Place the cuff on the upper arm and position it over the distal half of the biceps muscle (Figure 9.8). Secure it comfortably to the arm with the attachment device on the cuff. (Older cuffs wrap around the arm or have metal snaps; newer cuffs have Velcro closures.)

3. Place the stethoscope diaphragm at the antecubital fossa and over the brachial artery (Figure 9.9). Place the earpieces of the stethoscope into your ears. (Position them so they are comfortable.)

Figure 9.8
Placement of the Cuff.
Note the proper placement of the blood pressure cuff on the upper arm.

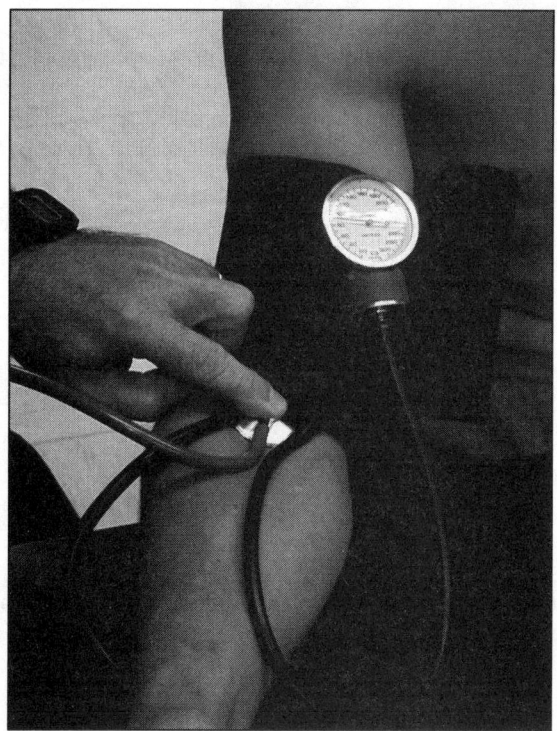

Figure 9.9
Measuring Blood Pressure.
Note the proper position of both persons in the measurement of blood pressure.

4. Begin pumping up the cuff, simultaneaously watching the gauge and the subject. Do not exceed 180 mm Hg in the cuff. (Pressures at or near this pressure are painful for most people. This pressure should be sufficient, however, to completely occlude the vessels in the upper arm.)

5. Slowly release the pressure in the cuff, at a rate of approximately 2 to 3 mm Hg per second. Listen carefully for any sounds.

6. As the pressure in the cuff drops and then equals the force being generated in the circulatory system, some blood squirts through the occluded vessel. The squirting blood makes a sound, which is referred to as a Korotkoff sound. (This is not the same as the heart sound.)

7. Note at which pressure the first sound is heard; this is the systolic pressure. The quality of the sound will change as the pressure drops in the cuff. The first sounds are sharp, distinct, and increase in intensity. This is followed by a period in which the sounds become muffled and decrease in intensity. Then, the sounds increase abruptly, and drop off abruptly as well.

8. The person reading the pressure notes when the last sound is heard; this is the diastolic pressure. (Knowing that the sound being heard is indeed the last sound is difficult. You will not know this is the last sound until you no longer hear any sounds. Therefore, you must watch the gauge carefully at this point in the recording.)

   First Determination:

   Systolic pressure: _____ mm Hg

   Diastolic pressure: _____ mm Hg

Determining Blood Pressure

9. Rapidly release the remaining pressure in the cuff, and allow the subject to relax and stretch his or her arm. It is recommended to repeat the measurement and compare the two readings to check your technique. (If the measurement will be repeated, allow the subject to rest his or her arm a minute to prevent pooling of blood in the lower arm, which would interfere with accurate readings.)

   Second Determination:

   Systolic pressure: _____ mm Hg

   Diastolic pressure: _____ mm Hg

   (If there is considerable variation between the first and second determinations, repeat the measurement a third time, and use the average of the two closest measurements.)

10. After everyone has determined their blood pressure, record all the pressures, listing all the systolic pressures together and all the diastolic pressures together. Then average these figures to see how close the figure compares to 120/80 mm Hg.

QUESTION: Based on how you just determined blood pressure, can you think of any other locations on the body at which you could determine blood pressure with this same method?

Observations:

Conclusion:

A person with normal blood pressure is said to have **normotension.** When the pressure meets established clinical criteria for chronic elevation, that person is said to have **hypertension** (*hyper* (G), over, above, excessive). Most of these people (95%) are referred to as having essential or primary hypertension, the cause of which is unknown. Consistently low blood pressure is called **hypotension** (*hypo* (G), under, beneath).

QUESTION: When taking the blood pressure, it is recommended to have the subject sit upright in a chair with the cuffed arm level with the heart. Why is this important to ensure an accurate reading?

Observations:

Conclusion:

---

148   SECTION 9   The Circulatory System

Journal Note Have you ever stood at attention as a band or a drill team member? Were you warned not to lock your knees? What happened if you locked your knees? Why?

Journal Note Many grocery, drug, and discount stores have automatic blood pressure devices. Try one. How does its measurement compare with your lab reading? Consider sources of error in this type of measurement. Can you change your reading by adjusting your position, moving, or clenching your fist?

Determining Pulse and Pulse Pressure

Everyone has felt their pulse sometime in their life. We do this by placing one or two fingers just above the wrist on the lateral forearm. We feel a movement, which actually is a change in pressure, of the blood passing through the radial artery on the inside of the wrist. This spot where we can feel our pulse is called a pressure point. Actually, there are several points on the body where fairly large vessels are palpable from the surface of the body. These are located on the forehead just anterior to the temple, behind the mandible along the throat, in the antecubital fossa, in the groin, behind the knee, and on the lateral surface of the ankle (Figure 9.10).

What exactly is the pulse? In other words, what is being felt by the fingers as we press various points on the body surface? It is the difference between the systolic and the diastolic pressure. When we push the pressure points, we are feeling the change

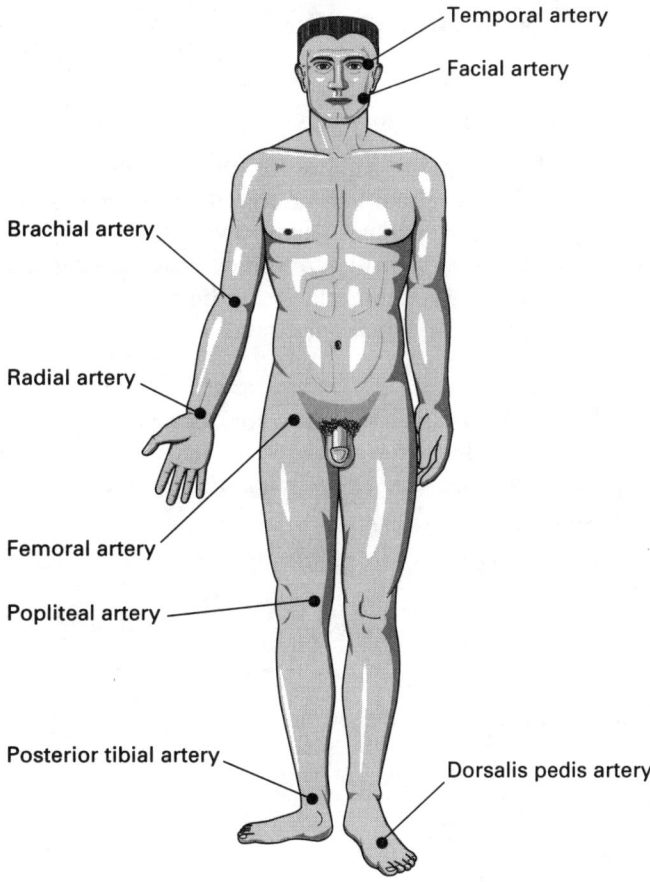

Figure 9.10
Pressure Points.
In addition to the lateral wrist location, the pulse may be palpated at numerous locations.

from high to low pressure. This is called the **pulse pressure.** If you just determined a blood pressure, calculate the pulse pressure using the following equation:

Pulse pressure = systolic pressure – diastolic pressure
Pulse pressure = _____ mm Hg

Taking the Pulse

Objectives:

- To understand the pulse and pulse pressure.
- To learn how to take the pulse by palpation.
- To appreciate changes in the pulse as a response to changes in body position.

Materials:

- Timing device.

Procedure:

1. While sitting or lying down, palpate a pressure point (most people use the pressure point at the wrist), and determine your pulse (or a partner's pulse) for 15 seconds.

   Pulse for 15 seconds: _____

2. Multiply this figure by four to get the pulse in beats per minute. (Pulse is always reported in beats per minute [bpm]).

   Resting pulse: _____ bpm

3. Stand quickly, and immediately record the pulse again.

   Sudden standing pulse for 15 seconds: _____

   × 4: _____

   Sudden standing pulse: _____ bpm

4. Collect the pulse readings for everyone in the clas,s keeping the resting values separate from the standing values. Plot the readings in an appropriate fashion to show how many individuals have a pulse within a certain range. Then, average the two groups of readings to determine the average resting pulse rate and the average rate after quickly standing. How do these values compare with the often-cited value of 72 bpm for a normal pulse rate?

QUESTION: When normal values are given for body parameters, these values are based on what? In other words, who were the individuals sampled to determine the normal values for pulse, blood pressure, blood sugar concentration, and so on? Or, phrased differently, what is a "normal" value?

Observations:

Conclusion:

5. If the class is large enough, subdivide the results for resting pulse into two groups: regular exercisers, or those who exercise at least three times a week for 20 to 30 minutes; and nonexercisers, or those who do not exercise regularly. Average these figures. Do these averages differ from the average that you just determined for the class as a whole?

Anatomy of the Heart and Vasculature

The heart probably is the most recognizable organ in the body (Figure 9.11). It is a double pump having a left and a right half, which are separated by the **interatrial** and **interventricular septum.**

Blood returns to the heart from the body via the superior and inferior vena cava. The superior vena cava actually drains a smaller volume of the body, but it is located superior to the heart and drains the upper chest, neck, head, arms, and shoulders. The inferior vena cava is the larger vessel, and it drains the lower body, including the abdomen, pelvic regions, and legs.

The blood flows from these vessels into the right atrium. An extension of the atrial chamber is an ear-like flap called the **auricle.** Blood flows from the right atrium into the right ventricle through the **right atrioventricular valve,** which also is known as the **tricuspid valve.** The right ventricle then pumps the blood into the **pulmonary trunk,** which supplies the lungs with the blood that needs to pick up oxygen and get rid of carbon dioxide. To enter this vessel, blood flows through the **right semilunar valve,** which also is known as the **pulmonary semilunar valve.** The pulmonary trunk branches into **left** and **right pulmonary arteries,** which carry blood to the lungs. Blood returns from the lungs through the **left** and **right pulmonary veins** to the left atrium. There also is an extension off the left atrial chamber called the **left auricle.** Blood flows from the left atrium through the **left atrioventricular valve,** or the **bicuspid valve.**

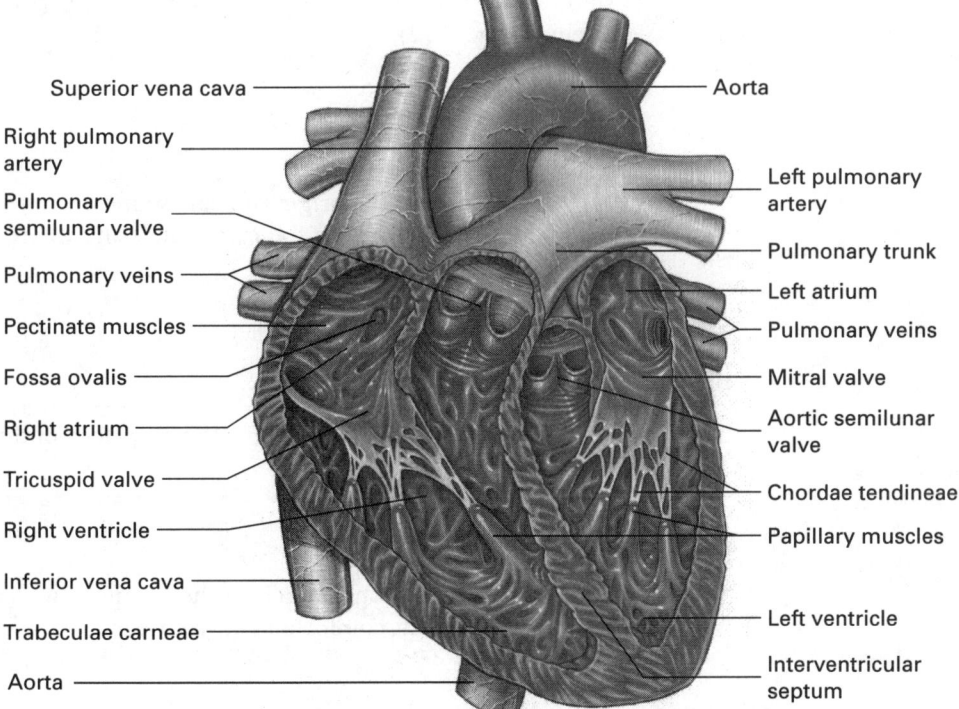

Figure 9.11
The Heart.
This frontal section of the heart shows the interior of the four chambers.

Anatomy of the Heart and Vasculature   **151**

Both the right and left atrioventricular valves require some reinforcement or strengthening to prevent their "blowing out" during the systolic phase. These valves are supported by the **chordae tendineae** and the **papillary muscles.** The chordae extend from the tips of the papillary muscles to the valve leaflets. As the heart pumps, the papillary muscles contract to hold these valves against the pressure that is generated by the heart.

Blood is pumped from the left ventricle into the major systemic vessel of the body, the **aorta.** To enter this vessel, blood goes through the **left semilunar** or the **aortic semilunar valve.** The aorta gives rise to the vessels that supply the head and upper body as well as becomes the thoracic aorta as it completes its sweeping arch and heads inferiorly. After the thoracic aorta passes through the diaphragm, it is referred to as the abdominal aorta.

Coming off the aorta immediately as it begins are the **left** and **right coronary arteries.** These vessels supply the heart muscle, and these are the vessels (or their branches) that become occluded and interfere with blood flow, possibly causing **angina** or **heart pain** or, if completely blocked, a **myocardial infarction** or **heart attack.**

The veins of the heart all drain into an enlarged blood space on the posterior surface, which is called the **coronary sinus.** This sinus drains into the right atrium.

Listening to the Heart

One of the first things a clinician likely will do during a routine checkup is listen to the heart. The stethoscope provides a means to hear the noises the heart makes and to judge if those sounds are normal. Auto mechanics also use stethoscopes to hear the slight noises an engine might be making that could indicate a problem. The only thing that makes noises in the heart, however, are the heart valves. Normal blood flow through the heart is nonturbulent, and therefore, silent. The **first heart sound** is caused by the two atrioventricular valves snapping shut. The **second heart sound** is caused by the semilunar valves closing. These two sounds are described as **lubb–dupp.** A trained person not only can hear these sounds but move the end of the stethoscope around to the various intercostal spaces and hear each valve more precisely (Figure 9.12).

Listening to the Heart

Objectives:

- To understand the origin of heart sounds.
- To learn how to listen to heart sounds and where on the chest to best hear them.

Materials:

- Stethoscope.
- Alcohol pad.

Caution: It is advisable to clean the earpieces of the stethoscope with an alcohol pad before use.

Procedure:

1. Obtain a stethoscope and clean the earpieces with an alcohol pad. Adjust the earpieces so they fit comfortably in your ears.

Figure 9.12
Listening to the Heart Valves. Shown are the locations on the chest at which the four different heart valves can be heard. The valves are not located directly beneath the marked locations, but each valve's sounds conducts best to the surface of the chest at one of these locations.

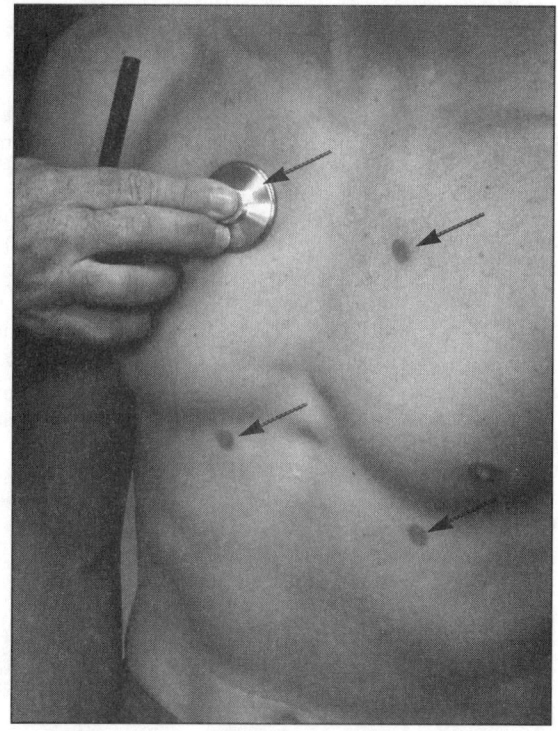

2. Using the stethoscope, listen to your heart or that of a partner. Try taking deep breaths. Does that change the sound? Now listen to the heart from the back. Do you hear any differences?

3. Referring to Figure 9.12, try to hear the different heart valves. It will help to find a quiet location, because background noise will interfere with hearing the valves.

QUESTION: If turbulent blood flow causes the Korotkoff sounds heard when determining blood pressure and the heart valves snapping shut causes the heart sounds, then what causes heart murmurs?

Observations:

Conclusion:

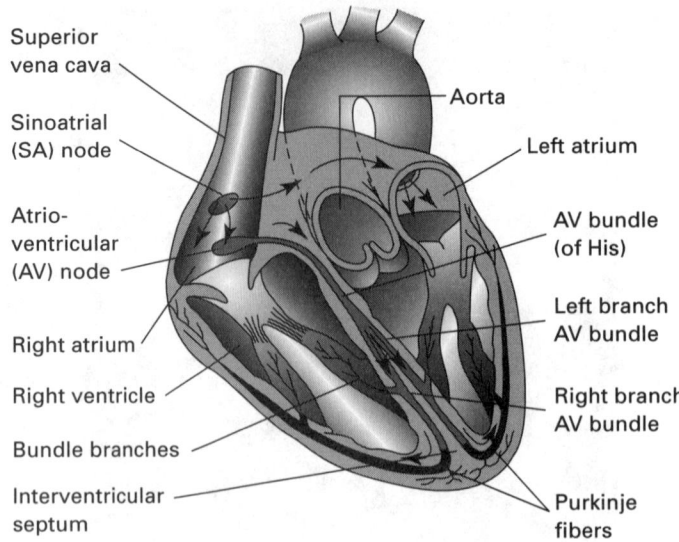

Figure 9.13
Conduction of the Heartbeat. This drawing of the heart shows the location of the pacemaker and the conduction system for the heartbeat.

Control of the Heartbeat

In healthy, normal individuals, the heart beats on its own. A structure called the pacemaker sends out a signal that causes the heart to beat. This tissue actually is in an area of specialized cardiac muscle tissue called the sinoatrial node (SA node) (Figure 9.13). It is located in the upper right atrial wall near the base of the superior vena cava, and it sends out its signal to the surrounding atrial muscle tissue, thus causing the atria to beat. The signal reaches another specialized area of modified cardiac muscle, which is called the **atrioventricular node (AV node),** in the lower right atrial wall. The signal causes this tissue to depolarize and, in turn, to send a signal that causes the ventricles to beat. This signal is carried to the base of the heart and causes the ventricles to contract. Running from the AV node is the **bundle of His,** or the **atrioventricular bundle,** which branches into two parts, the **left** and **right bundle branches.** Both the bundle and the bundle branches run through the interventricular septum, where they then branch into **Purkinje fibers,** which make the actual connection with the cardiac muscle cells of the ventricles. As the signal arrives, it causes these cells to contract.

Figure 9.14
Pacemakers.
Artificial pacemakers today are very small compared with the original, pocket-sized versions.

154   SECTION 9   The Circulatory System

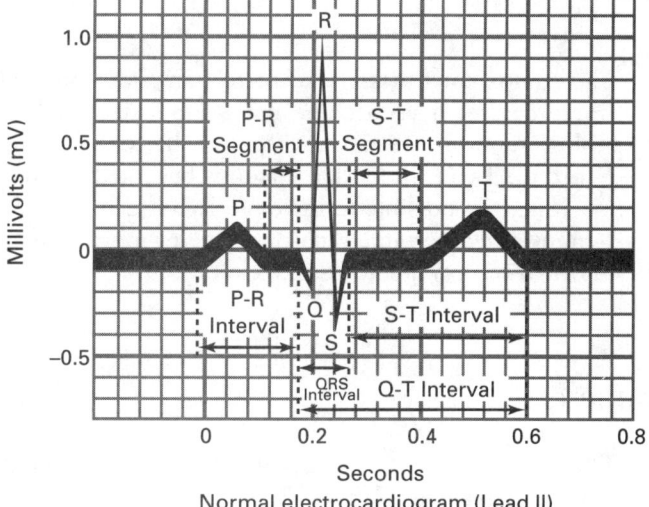

Figure 9.15
The ECG.
This is the typical pattern of waves recorded on an ECG: the P wave, the QRS complex, and the T wave.

You may have a relative or friend who, because of a birth defect or disease, does not send out a strong enough or a fast enough beat to control their heart. This might have been treated by insertion of an extracardiac pacemaker, which is an electronic device that sends an electrical signal to appropriate places in the heart, thus causing it to beat. Artificial pacemakers once were large, pocket-sized devices that would be placed outside the body, with wires running into the chest. Modern versions are so miniaturized and contain such long-lived batteries they are routinely implanted under the skin, where they can remain for long periods of time (Figure 9.14).

As the heart contracts, there is a tremendous amount of electrical activity in the heart muscle, which can be recorded on the surface of the body. Using electrodes on the surface of the body and proper filtering, the **electrocardiogram (ECG)** is recorded (Figure 9.15). There are numerous sets of leads, or ways to record the ECG. There can be three limb leads for a simple demonstration recording set-up; there are another three augmented limb leads as well. Augmented limb leads are those in which two leads are compared against the other one. A full ECG would require the six limb leads, and six more chest leads (Figure 9.16). In a typical ECG wave pattern, the **P wave** represents the electrical activity of the atria. The **QRS complex** is evidence of the electrical changes in the ventricles as they depolarize before contraction. The QRS complex is of higher amplitude, because the ventricles represent a much greater amount of muscle tissue. The **T wave** represents repolarization of the ventricles as they relax and refill. Much information can be gained from the ECG. It often is sent to a specialist (i.e., a cardiologist) for careful inspection and evaluation.

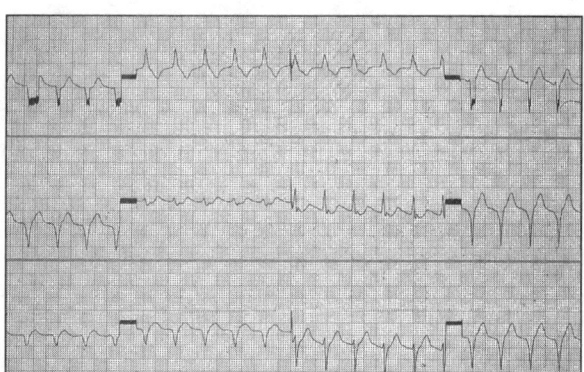

Figure 9.16
A full ECG record.
This is a typical record of a full, 12-lead ECG recording.

Control of the Heartbeat

Suggested Activity: ECG

The equipment necessary to record an ECG probably is not available in your lab. These devices have become quite sophisticated, however, and many are portable. Holter monitors are devices that can be worn by a patient for extended periods to record the ECG while that person is busy with his or her daily activities.

It may be possible to invite an ECG technician from a local clinic or hospital to demonstrate a portable recorder. One or more students could be hooked up to show a normal recording and its interpretation.

Blood and Blood Parameters

The fluid pumped by the circulatory system is called blood. Blood has a liquid portion called **plasma** (*plasma* (G), something molded) and a cellular portion made of red blood cells (RBCs), white blood cells (WBCs), and **platelets** or **thrombocytes** (*thrombo* (G), a clot). Red blood cells are discussed in the section on respiration, because these cells are associated with the respiratory function of the circulatory system. Other components of blood are dealt with here.

Plasma largely is water, with organic and inorganic substances added. The consistency and color of plasma largely is determined by the protein component, which contains albumin, plasma proteases, immunoglobulins, and hormones.

QUESTION: We all know what immediately happens as raw egg white (i.e., albumin) hits a hot frying pan. What do you think would happen to suddenly heated plasma? There is a special term that refers to the destructive effects of heat on proteins: denaturation. Do you think this process is reversible?

Observations:

Conclusion:

White blood cells also are called **leukocytes** (leuco (G), white), and they are the immune system components in the bloodstream. They help to protect us from the many infectious agents that constantly threaten our health, and they are divided into two major categories based on the presence of cytoplasmic granules: granulocytes, and agranulocytes (Figure 9.17). When stained, granulocytes have obvious granules in the cytoplasm. Agranulocytes have some cytoplasmic inclusions, but they do not show detectable granules when stained.

Three types of granulocytes commonly are identified. They are named based on their response to different types of stains:

1. **Neutrophils** are phagocytic cells with an obvious, dark-staining nucleus that often has many lobes. The cytoplasm is light colored, because the granules do not pick up much stain. Because of the multiple lobes of the nucleus, these cells may appear to have several nuclei, and they often are called polymorphonuclear leukocytes. In humans, neutrophils are the most common cell, comprising 60 to 70% of all WBCs.

**Figure 9.17
WBCs and Other Formed Elements.**
The cells shown are the formed elements found in the blood. They include red blood cells, WBCs, and the platelets.

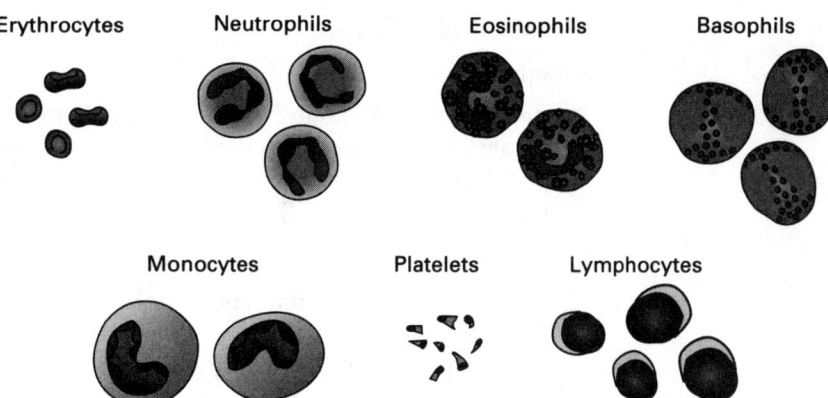

2. **Eosinophils** can be quite colorful depending on the staining process used. They also are thought to be phagocytic and are associated with allergies and parasites. They have tightly packed, uniformly sized, evenly distributed red or orange granules around an obvious nucleus. The nucleus also has lobes. This is not as obvious as in neutrophils, however, because of the darker granules. They are the next-to-least-plentiful cells, representing only 2 to 4% of all WBCs.

3. **Basophils** are characterized by nonuniform, randomly distributed, dark granules that actually may obscure the nucleus. These cells are associated with stress and parasitic infections. They are the rarest type of WBC, representing only 0.3 to 0.5%.

There are two different kinds of agranulocytes:

1. **Lymphocytes** are characterized by a rounded nucleus that can take up most of the cell; thus, there is little cytoplasm. The nucleus stains darkly, and the little cytoplasm evident appears to be smooth or agranular. There actually are large and small lymphocytes, with large lymphocytes having more cytoplasm. The cells referred to as B cells; T cells are specialized forms of these cells. Lymphocytes account for 20 to 30% of all WBCs.

2. **Monocytes** are the largest type of WBC, and they are phagocytic. Along with the lymphocytes, they represent a second player in the immune response. They have a large, rounded, lobed nucleus, with a fair amount of pale-colored cytoplasm evident. The nucleus is large and robust, not like the pinched lobes with narrow connections in a neutrophil. Usually, monocytes account for 3 to 8% of all WBCs.

Normally, there are 5 to 10×10^3 WBCs/μl. (One microliter is a very small volume, equal to 1/20 of a standard drop). The WBC count is an indicator of the general health of the body's systems, and different types of infection will manifest through elevated levels of different types of WBCs. A typical bacterial infection elevates the neutrophil level. Viral infections are difficult for the body to fight, and monocytes are associated with these. One of the classic signs of an acute infection is an elevated WBC count. Once an elevated count has been determined, it is informative to know which type of WBC is elevated. The number of WBCs per microliter is called the total WBC count; the count to determine the percentage of each cell type is called a differential WBC count. A total WBC count simply requires dilution and counting. The differential WBC count requires that cells be stained so they can be differentiated. Stains that normally are used include either Giemsa, which stains the WBCs different shades of purple, or Wright, which really is a mixture of stains that gives the cells shades of blue, purple, orange, and red. Wright stain is better during initial training to identify these cells, but once a person is trained, a black-and-white, good-contrast picture is adequate.

Differential WBC Count

Objectives:

- To understand the function of and be able to identify different types of WBCs.
- To gain experience in using oil-immersion lenses to observe the WBCs.
- To learn a technique, Wright staining, that allows the five types of WBCs to be differentiated (optional).

Materials:

- Microscope, with oil-immersion lens if available.
- Immersion oil.
- Prepared WBC smears, Wright-stained.

Procedure:

Note: This exercise on white blood cells can be carried out with prepared stained WBC smears or, if the supplies are available, you can make your own smears. Your instructor will let you know which procedure you will follow. If you will be making and staining your own smears, your instructor will give specific instructions for making smears and staining them.

---

1. Select a prepared blood smear, and place it on the stage of the microscope. Focus on the blood cells. An oil-immersion lens, though not essential to observing WBCs on a slide, will make identification easier. Switch through the different powers until ready to switch to the oil-immersion lens, if available. (If you are inexperienced with oil-immersion lenses, ask your instructor for assistance.) If an oil-immersion lens is not available, observe the cells at high dry (40 or 43× lens).

2. Place a drop of immersion oil on the slide, then switch the lens into place and focus.

Note: If using an oil-immersion lens, do not change the focus, even though it may appear that the lens will hit the slide. The lens must be in the oil for the system to work.

---

3. Make sure the area being observed appears to be one-cell deep. If the area in the field is not, find such an area on the slide.

4. Select a spot on the slide, preferably one near the edge, and begin identifying WBCs. Use Figure 9.17 for assistance. Move in only one direction until there are no more cells to count. Then, move the slide over one field, and continue counting. Make sure that no area is counted twice by following a consistent pattern in your counting (Figure 9.18).

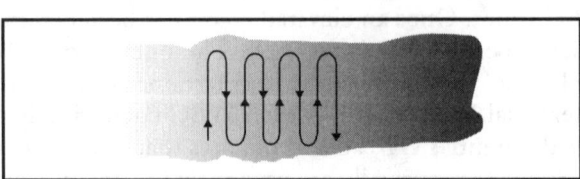

Figure 9.18
Counting WBCs.
To ensure the same WBC is not counted twice, a pattern of counting should be followed.

5. Keep a running tally of how many different types of WBCs you see. If you count at least 100 cells, then each number is equal to the percentage. It is recommended to count 200 cells, which will increase the chances of seeing a basophil and better represent the true distribution of WBCs.

Number of Cells Counted

    Neutrophils: _____

    Eosinophils: _____

    Basophils: _____

    Lymphocytes: _____

    Monocytes: _____

QUESTION: The slides you have just used are stained to make the WBCs more apparent and help you to differentiate the cells. You probably noticed the cells were colored differently. Can you describe how each type of WBC looks without describing its color? In other words, can you come up with a colorless description?

Observations:

Conclusion:

QUESTION: There are many more red blood cells (RBCs) than WBCs per microliter. When you count RBCs, how would the procedure be adjusted? Would you collect a larger or a smaller sample when counting RBCs as opposed to WBCs? Would you dilute the sample as much or more? The hemocytometer is a special slide used to count very small objects such as cells; if you count RBCs or WBCs, you would count them using a hemocytometer. Would you count as large an area or a smaller area for RBCs compared to WBCs when using a hemocytometer?

Observations:

Conclusion:

Journal Note No one likes to get sick, but we all do sooner or later. Sometimes, the sickness is simply a cold, the flu, or some other mild infection. Occasionally, the illness is a bit harder to ascertain. Or we may be just due for a checkup, and a blood test may be ordered. If you have a blood test performed during the semester, ask for a copy and study the printout. A typical report will have your specific results and often contain the normal range for all of the items tested. High or abnormal levels probably will be indicated as well. Make this a part of your journal. Your book, this manual, or your instructor may help to explain measurements on the sheet that you may not understand.

Section 10

The Digestive System

Note to the Student

The body is a complicated metabolic machine, and it requires the energy, as contained in the foods consumed, to operate its various parts. It is the digestive system that provides the proper conditions to break down foods into smaller "building block" components, which can be used either to build new parts, repair damaged or worn out components, or be burned as fuel to release energy. The digestive tract within your body really is a tube within a tube. The torso or trunk is a sealed tube (i.e., the body cavity), with the digestive tract being another tube that begins at the mouth and ends at the anus. Food in the tract is separated from the rest of the body. This is a protective feature that allows isolation of disease-causing organisms and undigestible or toxic foods. This separation is not always successful, however, because many organisms and consumed substances can cause life-threatening conditions even if technically outside of the body.

The digestive tract consists of the following structures: mouth, oral cavity, pharynx, esophagus, stomach, small and large intestines, rectum, and anus (Figure 10.1). Also associated with the tract are accessory glands of the pancreas, liver, and gallbladder. The basic histology of the digestive tract is fairly consistent throughout its 9.14-m (30-foot) length.

QUESTION: The actual length of the digestive tract is difficult to measure. In a living individual, the tract is moving, functioning, and cannot be exposed completely at one time. In a cadaver, the tract is relaxed and nonfunctioning. Do you think measures of the tract are longer or shorter in a cadaver versus a living individual?

Observations:

Conclusion:

Figure 10.1
The Digestive Tract.
This diagram of the digestive tract shows the major features associated with this system.

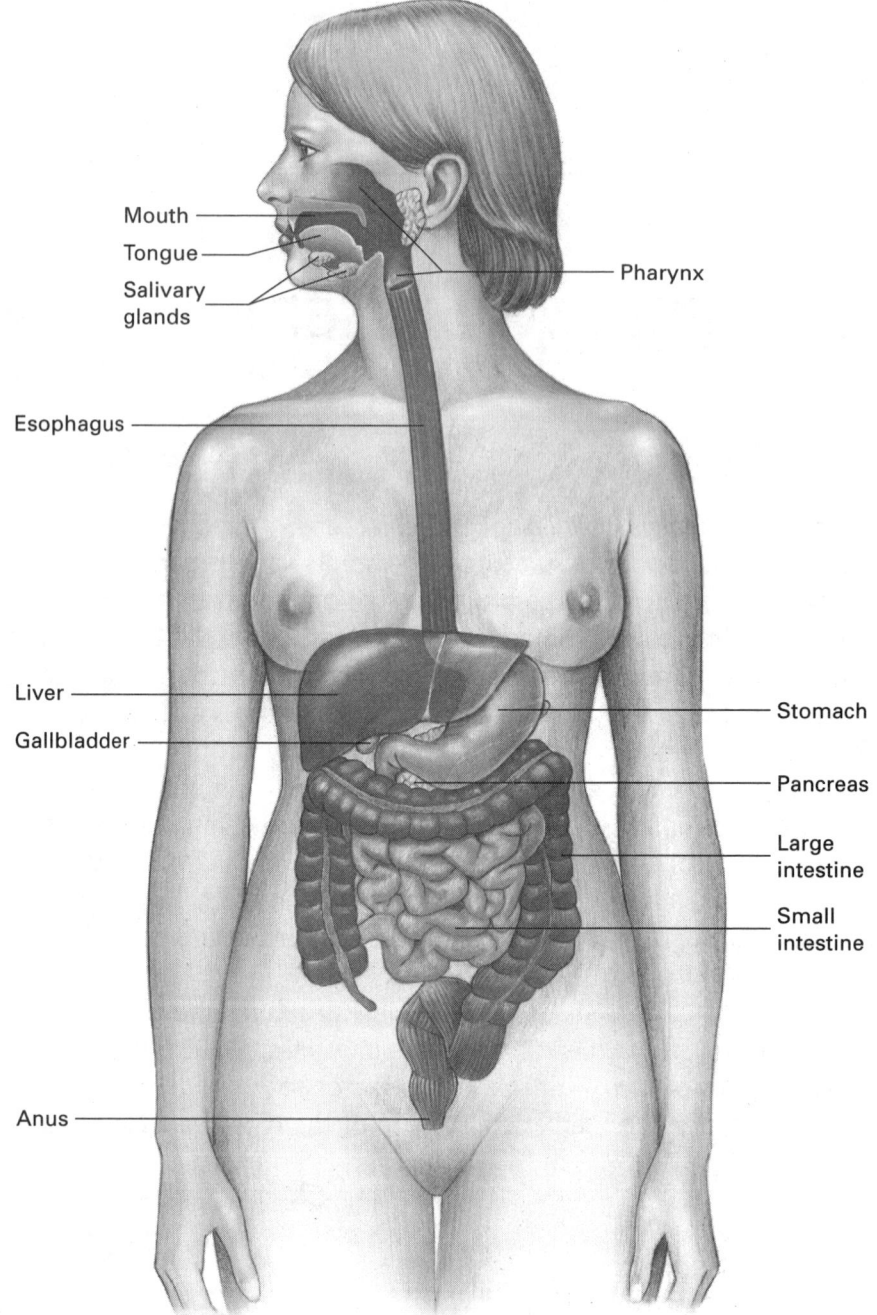

Four layers make up the wall of the digestive tract (Figure 10.2). The innermost layer is the mucosa, which is followed by the submucosa, the muscularis, and then the serosa. The **mucosa** essentially is simple columnar epithelial cells, with many goblet cells. The **submucosa** is rich in connective tissue and blood vessels, and it underlies the mucosa. Together, these two layers are folded in the stomach to produce the gastric pits/glands and, in the intestines, to produce the **villi.** Villi are finger-like folds of the wall that increase the surface area of the intestines, which provides enhanced absorption of digested foodstuffs. The villi gives the intestines a velvety appearance to the naked eye. The **muscularis** is the thick, smooth muscle layer of the intestinal wall, and it consists of an inner, circular smooth muscle layer and an outer, longitudinal smooth muscle layer. These two layers allow a change in both the diameter and the longitudinal dimension of the intestines. Working together, these layers are responsible for the mixing and propulsive movements of the gut.

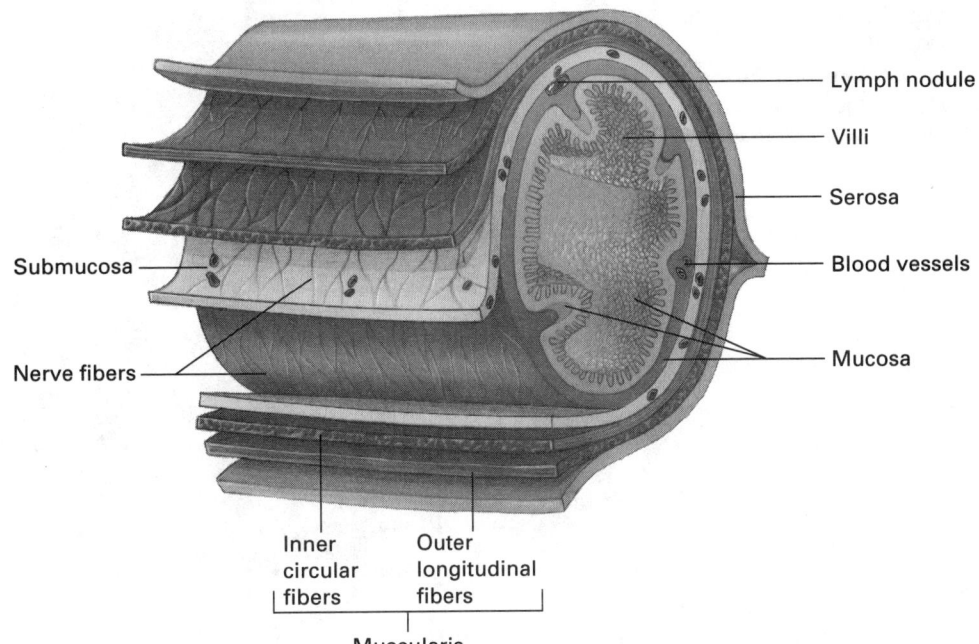

Figure 10.2
Cross-Section of the Digestive Tract.
This representative cross-section of the digestive tract shows the mucosa, submucosa, muscularis, and serosa layers. Note threre are modifications or specializations of these layers in various regions of the tract.

If you look at a slide of the digestive tract, it probably is a cross-section. The circular muscle will be cut with the fibers, but the longitudinal fibers will be seen in cross-section. The stomach also has a diagonal muscle layer, which further increases its potential motility. The muscularis' longitudinal layer in the large intestines is bundled into three bands, which give the large intestine a puckered appearance because the areas between the bundles pouch out.

The outermost layer is the **serosa.** This is the layer of connective tissue that surrounds the components of the tract and separates it from the other components in the abdominopelvic cavity.

Anatomy of the Digestive Tract

The opening that is the mouth is surrounded by the lips (Figure 10.3). The lips are composed of unkeratinized, stratified squamous epithelium, which allows the underlying blood vessels to show through and appear red. This part of the lip is called the **vermilion.** Passing through the mouth, the space containing the teeth and tongue is called the **oral** or **buccal cavity.** The mouth contains differently shaped teeth to handle different types of food. Humans are omnivorous, meaning they eat both animal and plant matter (Figure 10.4). The front four teeth are **incisors** for cutting food. Next are a pair of **canines** for holding and tearing, especially with animal tissues. Then come the **first** and **second premolars** and the **first, second,** and **third molars,** which are flat-surfaced teeth that crush and grind hard, tough plant and animal tissues. When food is swallowed, it enters the **oropharynx** as a wad or lump, which is referred to as a **bolus** (Figure 10.5). The tongue pushes both superiorly and posteriorly as the soft palate closes off the **nasopharynx** to prevent food from going into the nose. In the **laryngopharynx,** the muscles in the wall relax ahead of the bolus. To avoid choking caused by food entering the respiratory system, the **epiglottis,** which is the flap that covers the **glottis** (i.e., the opening for air through the larynx into the respiratory system), prevents food from entering the glottis.

Figure 10.3
The Mouth.
The mouth contains the tongue and teeth. It is bordered by the cheeks laterally and the palette superiorly.

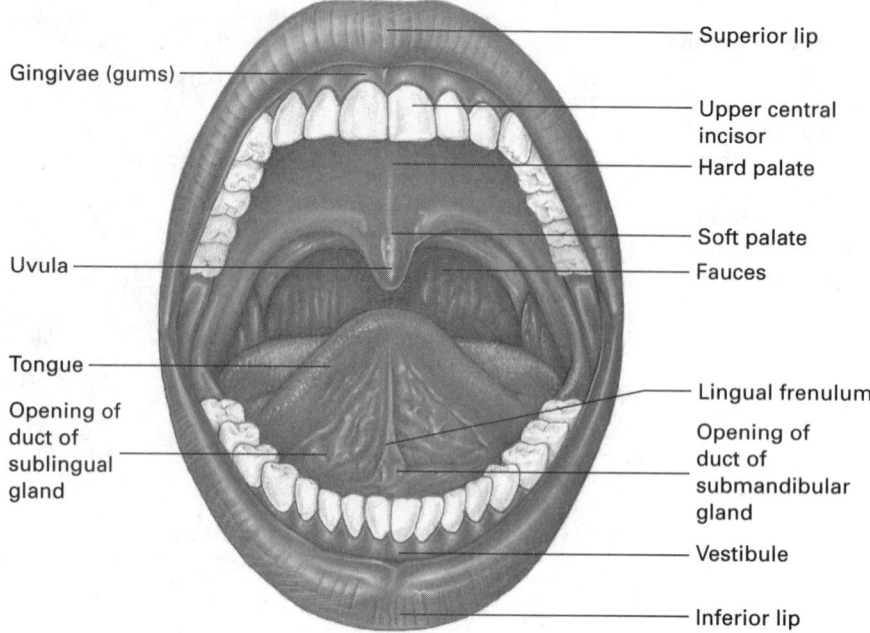

Figure 10.4
A Tooth.
This diagram of a tooth shows the major structural components.

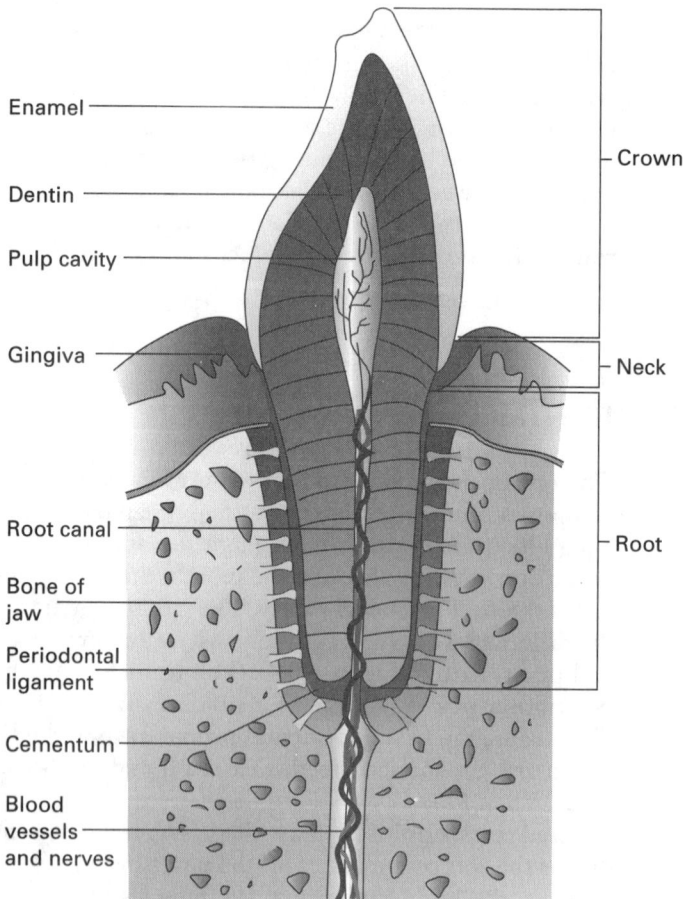

164 SECTION 10 The Digestive System

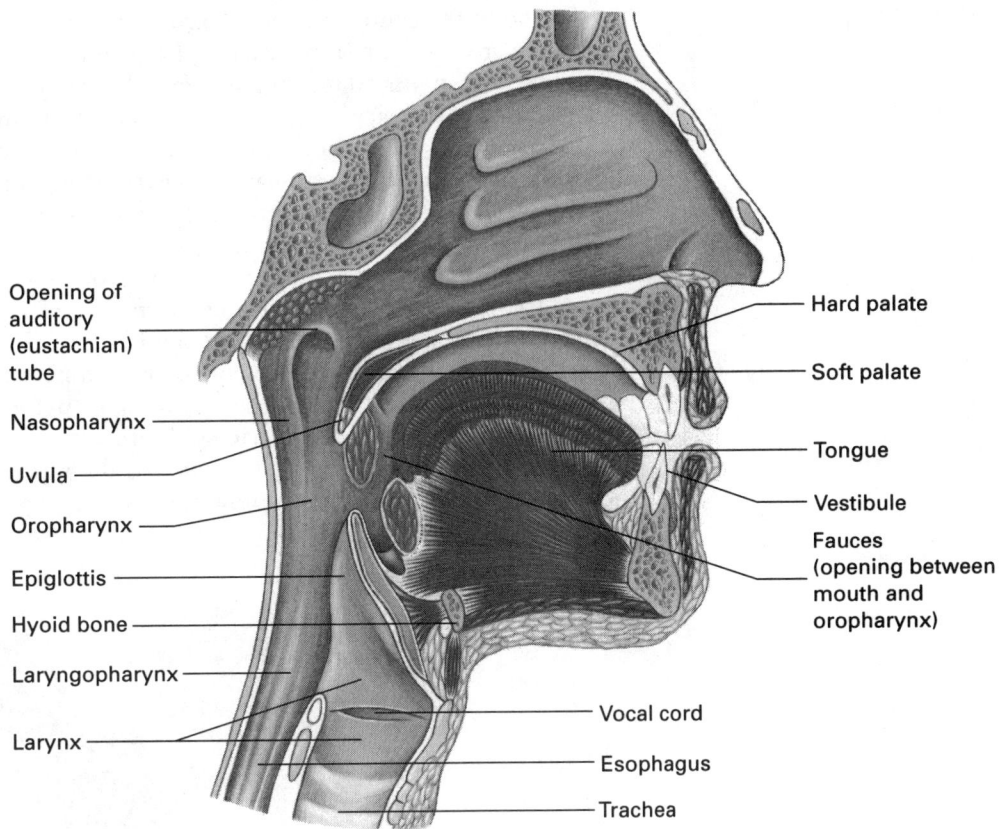

Figure 10.5
Sagittal Section of the Head.
This sagittal section of the head shows the mouth, pharynx, and upper esophagus.

QUESTION: Swallowing is a voluntary action. We place food in the mouth, chew, and then swallow. Smooth muscle is involuntary, and skeletal muscle is voluntary. What does this tell you about the nature of the muscle in the pharynx and the upper esophagus compared with that in the lower esophagus?

Observations:

Conclusion:

Anatomy of the Digestive Tract

The bolus then enters the **esophagus** for transit to the **stomach.** The esophagus is a muscular tube with a muscularis that contains voluntary skeletal muscle in its superior end changing to involuntary smooth muscle toward its inferior end. Food is pushed into the stomach, and the process of digestion, which started in the mouth with salivary amylase (an enzyme that digests starch) and chewing, continues under very acidic conditions. The stomach or gastric glands add the enzyme pepsin to digest proteins, acid to chemically destroy the food, and mucus to protect the stomach lining from acid attack and enzymatic digestion. The stomach is divided into an upper medial region called the **cardia,** or **cardiac stomach,** where the esophagus attaches (Figure 10.6). The uppermost, rounded portion is referred to as the **fundus,** or **fundic stomach.** The main portion is called the **body of the stomach.** Its lateral, longer edge is referred to as the **greater curvature** and its medial, shorter curve the **lesser curvature.** The body tapers into the **pylorus,** or **pyloric stomach,** which contains a valve or sphincter called the **pyloric sphincter.** It is the neuronal and hormonal regulation of the pyloric sphincter that controls the rate of digestion, and at this point, the liquefied food and secretions now are referred to as **chyme.**

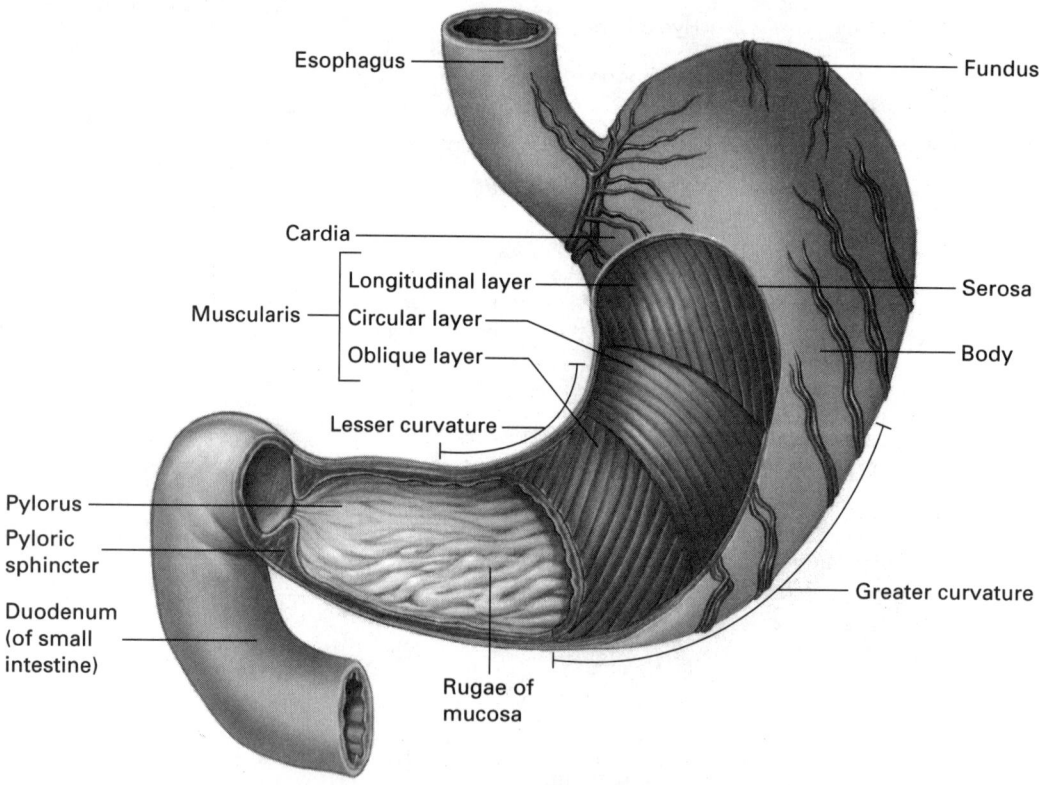

Figure 10.6
The Stomach.
This cutaway section of the stomach shows the layers of the stomach wall as well as its regions. Note the folds in the wall of the stomach, or the rugae, which allow the lining to stretch as the stomach expands with food.

The **small intestine** is the next part of the digestive tract, and it receives partially digested chyme from the stomach. The majority of digestion and absorption really occur in the small intestines. It also now is appropriate to talk about the **liver** and the **pancreas,** because they are directly involved in this part of the digestive tract. The small intestine is a tube, approximately 20-feet long, in which the chyme is mixed with many different digestive enzymes from both the pancreas and the intestinal glands. Enzymes break down the proteins, lipids, carbohydrates, and nucleic acids in the food, and then the smaller components are absorbed into the body. The pancreas is the major extraintestinal source of digestive enzymes, but it also is an important endocrine gland and is involved in the regulation of metabolism. The liver only produces one thing that is important for digestion: bile. Bile is an emulsifying agent that breaks fat into microscopic droplets, thus greatly increasing its surface area to speed lipid digestion. Bile is continuously produced by the liver.

QUESTION: Often, the most difficult material to clean off the dinner dishes is grease. Soap, however, seems to make grease disappear from the dirty dish water when it is added. Soap is a lot like bile. What does soap actually do to the grease to make it disappear?

Observations:

Conclusion:

When bile is not needed in the digestive tract, the flow of bile backs up into the gallbladder, which not only stores bile but also concentrates it by removing water. Some persons have their gallbladders removed in an operation called a **cholecystectomy.** Without adequate amounts of bile, however, lipids take much longer to digest and may cause digestive-tract discomfort or indigestion.

The pancreas is an important gland to the digestive system, and it also is part of the endocrine system. It produces many enzymes that assist in digestion of food but also bicarbonate to neutralize the acidic chyme that is flowing from the stomach.

The small intestine is made up of three segments. The **duodenum** is the 25-cm (10-inch) segment beyond the pyloric region of the stomach. Located in this segment is the opening of the **Ampulla of Vater** at the **duodenal papilla,** which is the combined ducts from the pancreas and the liver. The **jejunum** is the second segment, and it is very mobile and is approximately 2.5-m (8-feet) long. The **ileum** is the third segment, and it is the longest part of the intestines (3.6 m or 12 feet).

The **large intestine,** or **colon,** is the final segment of the digestive tract (Figure 10.7). It is approximately 1.5-m (5-feet) long and is composed of several segments. The portion at which the small intestine connects to the large intestine is the beginning of the **ascending colon.** The **ileocecal valve** controls the flow of material from the small to the large intestine, and a pouch called the **cecum** extends inferiorly from this point of attachment. A worm-like extension off the cecum is called the **vermiform appendix;** it is thought to be a vestigial structure containing some lymphatic tissue in humans and may be a remnant of a much larger cecum found in other mammals. The longitudinal muscles in the wall of the large intestines are grouped together into bands, which are called the **taenia coli.** This allows, or causes, the wall of the colon to form pouches, known as **haustra** (singular, *haustrum*).

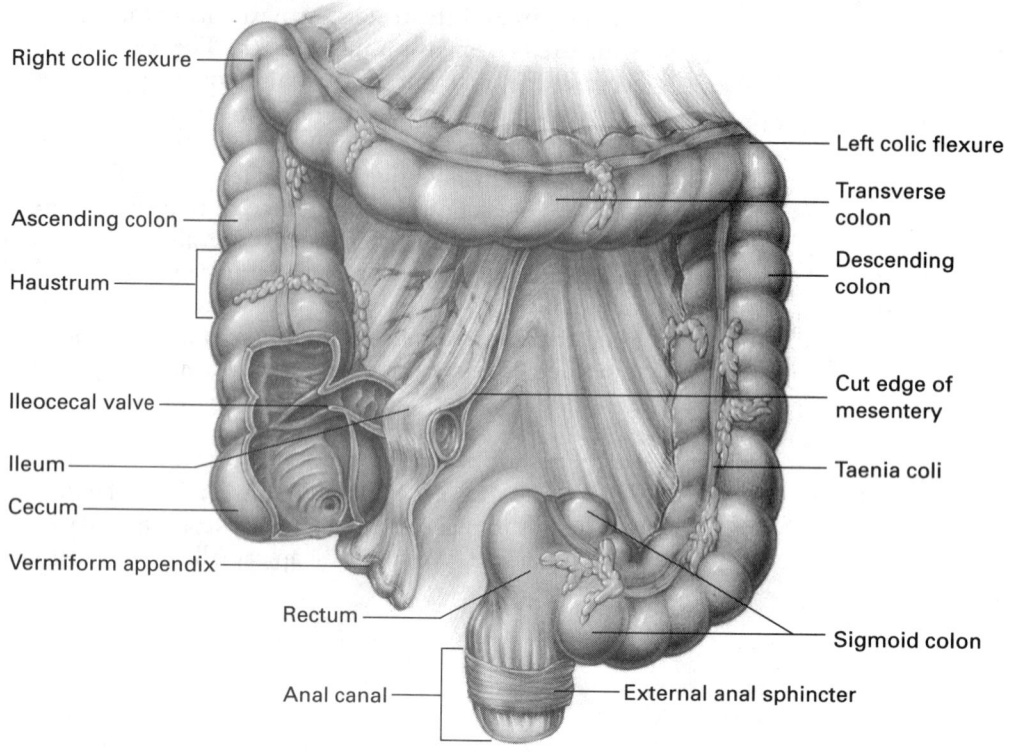

Figure 10.7
The Large Intestines.
This diagram of the large intestines shows its major segments. Note the pouches, or haustra, in the wall, which are caused by the bands of longitudinal muscle.

The next segment of the colon is called the **transverse colon.** It traverses from the end of the ascending colon just inferior to the liver and along the right side of the abdominal cavity, crossing to the left side of the abdomen, where it connects to the **descending colon.** This passes inferiorly and becomes the **sigmoid colon,** which is so named because of the curve it takes to align with the midline. The last portion of the colon is the **rectum,** or the chamber from which feces are evacuated from the body during defecation. The anus is kept closed by both the **internal** and **external anal sphincters.** The internal sphincter is composed of smooth muscle and is involuntary; the external anal sphincter is composed of skeletal muscle and is voluntary.

QUESTION: What muscle does a baby need to learn how to control to become "potty-trained"? Do you think a comparable set of muscles controls the flow of urine from the bladder?

Observations:

Conclusion:

Journal Note People vary regarding their regularity. Regularity refers to the frequency and consistency of defecation. Some have bowel movements each day and at a certain time of day. Others have bowel movements infrequently. Older individuals sometimes become obsessed with regular movements and resort to laxatives to stimulate defecation. How regular are you? Have you ever used a laxative? How do laxatives work? (Keep a personal diary of your bowel activity. It can be useful when trying to diagnose some disease states.)

Enzymes

Enzymes are proteins that serve as biological catalysts and, as mentioned, are important in digestion. A catalyst is an agent that facilitates or helps out in a reaction. It is not used up in the reaction so it remains to carry out the same reaction numerous times. Neither catalysts nor enzymes, however, can make an impossible reaction occur.

When discussing enzymatic activity, we call the substance that will react the substrate and the substance produced in the reaction the product. We express the reaction sequence as follows:

$$S_1 + S_2 \rightarrow E S_1 S_2 \rightarrow P + E \quad \text{or} \quad S + E \rightarrow ES \rightarrow P + E$$
$$S + E \rightarrow ES \rightarrow P_1 + P_2 + E$$

Only a single substrate (S) is shown the first equation, but several substrates can react to produce a single product (P). Alternatively, a single substrate molecule can be cut or cleaved to produce several product molecules. Note also in the equation that the enzyme (E) and S unite to form an Enzyme-Substrate (ES) complex as part of the reaction mechanism. Note also, however, that the S becomes a P but that the E remains the same.

All reactions require the addition of energy to make the reaction occur. In spontaneous reactions, all the added energy, and some of the energy already present in the substrate, also is released, and it is this loss of energy that makes the reaction spontaneous. The term that refers to the action of an enzyme is its **activity.** Enzymes facilitate reactions by helping the substrate get in the right position so that the reaction can occur under more favorable circumstances. Because of these favorable circumstances, the reaction can occur at a lower energy; therefore, we say that enzymes lower the activation energy of the reaction. Activation energy is the energy that must be added to the substrate for the reaction to occur. Thus, if the activation energy is indeed lower, then the reaction should occur more readily and, therefore, faster. Thus, enzymes speed up reactions. It generally is stated that enzymes speed up the rate of a reaction by 1 million–fold. If enzymes are biological catalysts, then living things must use enzymes. If there were no enzymes, then organisms would have to operate 1 million–fold slower than they can with enzymes. Therefore, living things could not exist at temperatures that are compatible with life.

To model how enzymes facilitate reactions, researchers have proposed several ways of imagining their workings. Because enzymes are proteins, they have the same characteristics as those associated with proteins, which are comparatively large molecules that have a distinct shape. The interaction between substrate and enzyme proposed initially was that of a key (i.e., substrate) fitting into a lock (i.e., enzyme). This view helped to explain several features of enzyme activity, but the model had a few weaknesses. The current analogy is referred to as an "induced-fit model." This implies that the enzyme and substrate fit together, but with the enzyme enveloping or conforming to the substrate. This view lacks the rigidity of the "lock-and-key model" but still focuses on the shape of the components. Thus, because enzyme and substrate shapes are so important and enzymes affect reaction-energy levels, anything that alters the shape of an enzyme (i.e., pH) or the energy level of the environment (i.e., temperature) should affect enzyme activity.

Inorganic Catalysis versus Enzymatic Catalysis

To show the drastic difference between a chemical catalyst versus an enzyme in facilitating a reaction, we observe the conversion of hydrogen peroxide into water and oxygen. The reaction is as follows:

$$2 H_2O_2 \rightarrow 2 H_2O + O_2$$

Iron oxide (i.e., rust) serves as an inorganic catalyst for this reaction. The enzyme catalase also facilitates the same reaction. Many body tissues have catalase to protect them from the harmful effects of hydrogen peroxide, which is produced as a hazardous waste material by many reactions in the body.

Degradation of Hydrogen Peroxide

Objectives:

- To appreciate the difference between an inorganic catalyst and a biological cata lyst (i.e., an enzyme).
- To develop skills for observing a reaction.

Materials:

- Test tubes and test-tube racks.
- Rusty nails.
- Iron oxide powder.
- Beef or chicken liver.
- Hydrogen peroxide, fresh 3% (household strength).

Procedure:

1. Secure a large test tube, and measure 2 ml of hydrogen peroxide into the tube. Add a rusty nail to the tube. Observe.

2. Obtain another test tube, and add 2 ml of hydrogen peroxide to the tube. To this tube, add a pinch of iron oxide powder. Observe. What happens when you shake the tube?

QUESTION: Assuming a change in the reaction rate occurred when you shook the tube, what important physical and biological principle does this illustrate?

Observations:

Conclusion:

3. Obtain another large test tube, and add 2 ml of hydrogen peroxide to the tube. To this tube, add a 0.5-cm^2 piece of fresh liver. (Liver tissue has high levels of catalase present; any readily available fresh or thawed liver will work.). Observe.

4. Grade the rates of reaction on some scale (e.g., 1–10) for the three different tubes. Feel the last tube. What does this observation tell you about the reaction?

QUESTION: If the liver caused an increased reaction rate, what could you do to the liver, based on the previous question, that would further increase the reaction rate?

Observations:

Conclusion:

Question: What effect would thoroughly cooking the liver have on its ability to break down hydrogen peroxide?

Observations:

Conclusion:

Enzymatic Activity

Enzymes can be purchased as dried, purified powders from chemical supply houses, or they can be collected from living sources, such as plant and animal tissues. These, however, typically are crude extracts that contain many enzymes. Depending on the particular enzyme and source, they also can be quite expensive. Our source of enzymes will be human saliva.

Human saliva contains only one enzyme, salivary amylase, which breaks amylose or starch into the disaccharide (*sacchar* (G), sugar) maltose. (Enzymes are named by taking the substrate name or the chemical removed by the enzyme and adding the suffix *-ase* to it. Hence, amylase is the enzyme that breaks down amylose into smaller components. Older names for enzymes did not follow this rule.) The reaction of amylase breaking down amylose into maltose can be followed colorimetrically. Iodine will react with starch to give a blue or purple color; iodine plus maltose does not produce any color. Therefore, when iodine is added to the substrate, a blue/purple color is attained early in the reaction. As the reaction proceeds, the color lightens and takes on some reddish hues resulting from intermediates of this reaction that are called **erythrodextrins** (*erythro* (G), red). Finally, when the reaction is complete and all the amylose and erythrodextrins are broken down into maltose, the only color present is that of the iodine solution itself. This point is called the achromatic point (*a* (G), without; *chrom* (G), color).

We now investigate the effects of temperature and pH. The following temperatures are tested: ice bath, room and body temperatures. There also are certain precautions to be aware of to successfully complete this exercise. For example, iodine will stop or "kill" the reaction. Use care when adding iodine and starch-enzyme mixture to the spot plates. Also, do not touch the tips of either dropper to the surface of the spot plate, because this might contaminate either or both solutions.

Salivary Amylase Activity: pH and Temperature Effects

Objectives:

- To study an enzyme (i.e., salivary amylase) that occurs in human saliva.
- To quantify an enzymatic reaction rate by visualizing the stages of a reaction and its end point.
- To appreciate some of the factors (i.e., pH and temperature) that affect enzyme activity.

Materials:

- Human saliva, diluted 1:4 with 10-mM NaCl solution.
- Test tubes and test-tube racks.
- Porcelain/plastic spot plates.
- Eye droppers (Pasteur pipettes with long tips).
- Starch solution, 1%.
- Iodine solution, 2%, in brown dropper bottles.
- Buffer solutions at pH 5, 6, 7, and 8.
- Water baths or insulated buckets.
- Ice.
- Graph paper for plotting results.
- Glass marker.

Procedure:

Caution: Because saliva is a body fluid, exercise care in its handling and disposal. Wear gloves when handling saliva samples, avoid spilling saliva, and wipe up any spills that occur. Dispose of saliva as instructed. Clean glassware thoroughly, or place in the special wash containers provided.

---

1. Secure a test-tube rack and nine large test tubes that will readily hold 10 ml (Figure 10.8). In each tube, place 5 ml of 1% starch solution. In the series of tubes to be used to study pH effects, add 4 ml of the appropriate buffer. (Suggested pH buffers are 5, 6, 7, and 8.) Label the tubes with the buffer contained and with your group number. Allow these tubes to reach room temperature. In the series of tubes for studying temperature effects, add 4 ml of the pH-7 buffer to the starch already in the tubes. Place these tubes in the different water baths or ice bath to reach the desired temperature. Suggested temperatures to test are: ice, room, 37°C, 45°C, and 65°C. Label the tubes according to temperature and your group number.

Note: There are two approaches for obtaining the saliva solution. Saliva typically is too concentrated when obtained directly from the mouth; that is why it must be diluted. Your instructor may collect the saliva or instruct a volunteer to do so. Large amounts of saliva can be obtained by chewing on a clean rubber band or piece of wax, or every group in the class can collect their own sample. The former method is preferable, because it allows all groups to compare their results to those of the others in the class. If every group uses a different saliva solution, comparisons are meaningless.

Figure 10.8
Set-Up for Enzyme Activity Assay.
Shown are the various items necessary for assaying the activity of salivary amylase. Note the large test tubes, long Pasteur pipettes, and spot plate.

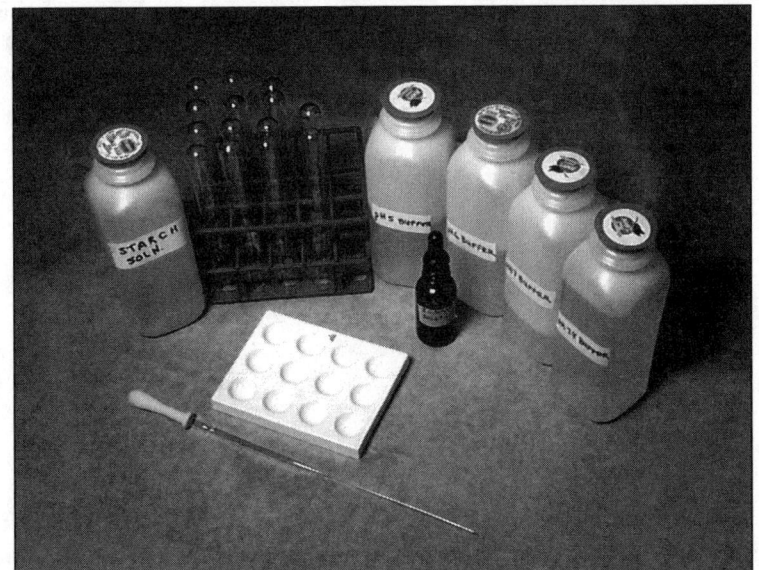

2. Secure a porcelain/plastic spot plate, brown glass bottle with iodine solution, Pasteur pipette, and rubber bulb for the pipette. When ready to analyze the first tube, add 1 ml of the saliva solution to that tube and begin timing. Be sure the starch, buffer, and saliva solution are thoroughly mixed by using the Pasteur pipette to suck up and squirt out the mixture several times.

3. At 30-second intervals, remove two or three drops of the mixture and place these in an empty hole on the spot plate. Add two or three drops of the iodine solution to the mixture on the spot plate, and ascertain the color. Swirl the plate to mix. Continue taking samples at 30-second intervals until the only color upon iodine addition is that of the iodine. At this point, stop and record the achromatic time. Be sure not get any iodine into the reaction tube, because this will terminate the reaction. Also, do not touch any of the droppers to the surface of the spot plate, because this may contaminate the stock bottles.

4. Continue with the other tubes until both series are complete. It may be necessary to repeat the first tube you assayed, because your initial inexperience may have affected these results.

5. When presenting your results, you should see that of the chosen temperatures and pH values, the best temperature and best pH were not at either extreme of the range. Therefore, if you plot these results, the curve will be U-shaped. This, however, means that the optimum pH and optimum temperature for this enzyme is at the bottom of the curve.

6. It is better to plot your results after some minor manipulation. Rather than plotting the pH or temperature against the achromatic times, convert the achromatic times to the inverse, or 1/achromatic, time. This will invert the plots and put the optimum pH and optimum temperature at the peaks of the curves.

Enzymatic Activity

QUESTION: Based on your results, what happens to enzyme activity when enzymes are stored at cold temperatures? What happens to enzyme activity when food in your mouth is swallowed into the very acidic stomach?

Observations:

Conclusion:

QUESTION: Food is cooked for several reasons. Many foods are more digestible and tasteful when cooked. Cooking also kills harmful organisms that may be on or in the food. Based on your results with temperature effects on enzymes, what will cooking do to enzyme activity?

Observations:

Conclusion:

Journal Note Some people are lactose intolerant. Some get indigestion from certain types of food, and certain foods cause individuals to have a lot of gas or flatulence. How do these observations relate to the presence or absence of different enzymes? Are dietary supplements available to help these people?

Nutritional Analysis

Food supplies the materials we need to build new tissues and to repair or replace parts that are damaged. Each day, you lose 100 billion cells from your body. Most of these are red blood cells that have worn out and cells that are lost from the epithelium making up your epidermis and the mucosa of your gut. These cells must be replaced just to maintain a status quo. If we still are actively growing, we also need nutrients to provide for additional tissues.

Nutritional research is difficult, because even when using experimental animals, there are distinct ethical considerations. To determine if a particular nutrient is necessary for development or growth, an experimental animal often is fed a diet that is deficient or lacking in that particular constituent. The consequences of this may be developmental deformities or death. This is an extreme way to determine if a nutrient truly is necessary, and it highlights why these studies are impossible to perform with humans. To deprive a boy or girl of some nutrient is both ethically and morally wrong. Therefore, most of our knowledge about nutritional needs is based on the results of animal experimentation and serendipitous discovery.

Nutritional Needs

Our nutritional needs typically are broken down into organic substances, minerals, and vitamins. When discussing nutrition, there are two important terms: essential, and limiting. In the nutritional usage, essential refers to nutrients that our body cannot produce and, therefore, must be consumed. Limiting refers to nutrients that our body can make, but not in the quantities necessary for normal growth and development; therefore, a dietary source is recommended. We need essential amino acids, a carbohydrate source, essential lipids, nucleic acids, minerals, and vitamins. Nucleic acids are present in any animal or plant source material, because all of these contain cells, which all have a nucleus with DNA and RNA. Our body does not absolutely need a specific carbohydrate, however, because these compounds are very interconvertable. If a person has a carbohydrate-deficient diet, the body can synthesize needed carbohydrates if there are extra proteins and lipids. Even so, carbohydrate-deficient diets are not recommended. A ready supply of carbohydrates guarantees that the brain and other tissues have adequate, easily metabolizable energy sources.

QUESTION: Sylvester Stallone trained extensively for the Rocky movies. This involved weight lifting and "bulking up" by eating lots of proteins. Stallone stated that he felt mentally slow during this period. Based on his comments, what part of the body probably was being deprived of a necessary nutrient? What class of nutrient may have been lacking in Stallone's diet?

Observations:

Conclusion:

There is need for a word of caution here. Many unproven claims have been made regarding the benefits of various dietary supplements. If you are going to supplement your diet, for whatever reason, investigate and obtain sound information from trained professionals. Moderation is a good rule to follow wherever health is concerned.

Nutritional Analysis

Objectives:

- To better appreciate the different food items needed for adequate nutrition.
- To become aware of the current condition of an individual's food intake regarding caloric and nutritional content.
- To use book sources or computer software to analyze food-consumption data.
- To keep thorough records of food intake over the course of the exercise.

Materials:

- Food intake record sheets (student- or instructor-prepared).
- Nutritional analysis handbooks or computer software for analyzing diet.

Procedure:

1. Your instructor will provide a nutritional analysis book or computer program. Familiarize yourself with the way in which foods are listed in this particular book or program.

> Note: Numerous shareware programs that are very affordable as well as software that is commercially available will provide an easy way to introduce computers into this course. (Shareware is the term for thousands of inexpensive—or free—software that is available from agencies that specialize in this type of software or companies that sell these programs for a nominal fee.) There even are free resources on the World Wide Web. One easily accessible program for nutritional analysis is at http://www.ag.uiuc.edu/~food-lab/nat/. (Do not include the last period.) The "nat" stands for Nutritional Analysis Tool. Most students will find their dietary analysis eye-opening as they see how poor their diets compare to nutritional recommendations.

2. The most difficult part of this nutritional analysis will be to accurately estimate the amounts of the foods you are eating. Familiarize yourself with the units used to measure food consumed.

3. Record everything you eat for meals and snacks over three days. Pick typical days, and try to avoid any holidays or weekends, when your intake would not be typical. Remember to estimate serving sizes when recording intake. If eating fast food, ask the restaurant for portion-size and nutritional information for your estimations.

4. After the three-day diet has been recorded, analyze your results. Analyze your intake on a daily basis, and then average the results for the three days. How close are you to meeting nutritional guidelines? Use the U.S. Department of Agriculture guidelines provided in Table 10.1 for comparison. Is your intake erratic, with quantities being over the limits on one day and under the limits on another? How is your vitamin intake? Are you eating as many calories as you thought, or are you eating more than you realized?

QUESTION: Many people pay a lot of money for nutritional supplements and eat specific foodstuffs or ingredients, because they know—or think—their body needs them. What happens to foodstuffs in the digestive tract? Do they stay intact and get absorbed whole so that a specific food will get directly into the body?

Observations:

Conclusion:

Table 10.1 U.S. Department of Agriculture Recommended Dietary Allowances (RDA) Regarding Daily Intake for Proper Nutrition.

Lipids

Because lipids are so plentiful in foodstuffs, there is little concern regarding adequate lipids being in a diet. The fatty acids **linoleic acid** and **linolenic acid,** however, cannot be made and must be consumed. These are polyunsaturated fatty acids, and they can be obtained from fish, safflower oil, sunflower oil, canola oil, and corn oil. It is recommended that no more than 30% of the calories in a diet come from lipids.

Proteins

There are 10 essential amino acids: **phenylalanine, valine, tryptophan, threonine, isoleucine, methionine, histidine, arginine, leucine,** and **lysine.** The term complete protein refers to compounds that contain all the essential amino acids. These are available in meat, fish, poultry, cheese, and eggs. Plant sources of protein are incomplete; therefore consumption of a variety of different plant sources is required for strict vegetarians to obtain adequate essential amino acids.

Carbohydrates

Most sources of carbohydrates are plant sources. Sucrose from sugar cane and sugar beets, simple sugars in vegetables, fruits, and honey, and starch from vegetables, fruits, and grains represent the variety of carbohydrates that are available. There are no essential carbohydrates.

Vitamins

Water-soluble vitamins
Thiamine (vitamin B^1)—unrefined whole grains and legumes, pork: 1.2–1.5 mg RDA.
Niacin (nicotinic acid)—lean meats, liver, peanuts, yeast, cereal bran and germ: 15–10 mg RDA.
Riboflavin (vitamin B^2)—dairy products, eggs, liver, leafy green vegetables: 1.1–1.8 mg RDA.
Pantothenic acid (vitamin B^3)—liver, egg yolks, wheat bran, and fresh vegetables: 5–10 mg RDA.
Biotin—dairy products, liver, egg yolks, yeast, and some produced by bacteria in intestines: 0.15–0.30 mg RDA.
Folic acid—leafy green vegetables, liver, lima beans, asparagus, whole grains, nuts, legumes, and yeast: 0.05 mg RDA.
Cobalamin (vitamin B^{12})—animal origin foods: 0.003 mg RDA.
Pyridoxine (vitamin B^6)—yeast, wheat, corn, egg yolks, liver, and meat: 2 mg RDA.
Ascorbic Acid (vitamin C)—citrus fruits, parsley, and tomatoes: 45–80 mg RDA.
Fat-soluble vitamins
Vitamin A—dairy products, liver, egg yolks, yellow and green vegetables: 1000 retinol units (5000 IU) RDA.
Vitamin D—fish, eggs, liver, and exposure to sun: 300–400 IU RDA.
Vitamin E—wheat germ and vegetable oils, liver, dairy products, and leafy vegetables: 12–15 mg RDA.
Vitamin K—intestinal bacteria: unknown RDA.

Minerals

Certain elements needed in significant quantities each day:
Sodium, potassium, chlorine, magnesium, and sulfur.
Some of the trace elements needed by the body:
Iron, calcium, phosphorus, iodine, copper, zinc, manganese, chromium, and selenium.

Methods for Determining Body Composition

During the past 10 years, there has been considerable interest in health- and wellness-related issues. It is partly a result of the strong associations of cigarette smoking with numerous cancers and cardiovascular and respiratory conditions, and partly a result of the links between diet and cancer, atherosclerosis, coronary artery disease, heart attacks, and strokes. Efforts to control or contain the increasing cost of health care has focused attention on preventive measures as well.

One factor that can be monitored to gauge the overall success of a fitness or weight-loss program is determining the percentage body fat. Several techniques allow this parameter to be measured. The gold standard method to which other methods are compared is underwater weighing (Figure 10.9), which involves determining a person's weight in air and then underwater. The premise of this method is that fat, which is less dense than water (because it contains less water), will cause the "fatter" person to weigh proportionately less underwater compared with his or her weight in air. In this technique, certain assumptions are made, and the body's density is determined. From that density, the percentage fat is calculated. This method is cumbersome, expensive, and not readily transportable. There are numerous other methods, but most are either highly technical, very expensive, or both.

Figure 10.9
Underwater Weighing.
This method, which involves weighing a person in the air and then underwater, is the basis on which other methods of percentage body fat determination are compared.

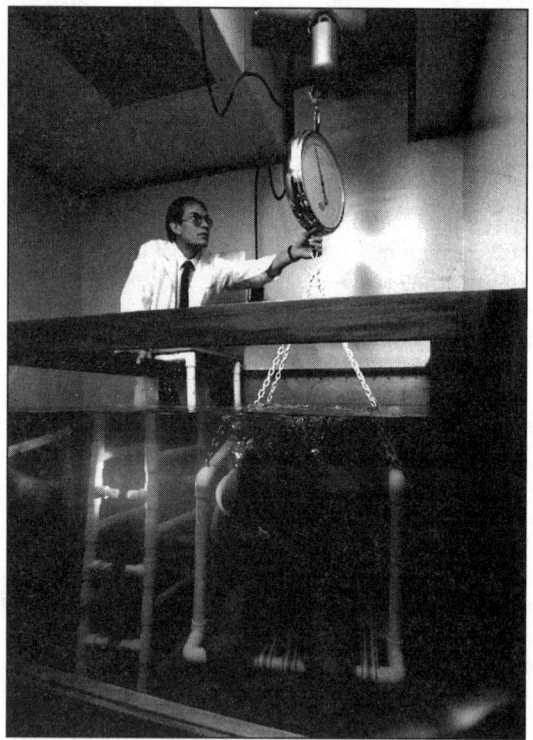

Figure 10.10
Skinfold Calipers.
Shown is Fat-O-Meter, a very affordable set of calipers that are available for percentage body fat determination from Novel Products, Inc. (P.O. Box 408, Rockton, IL 61072-0408).

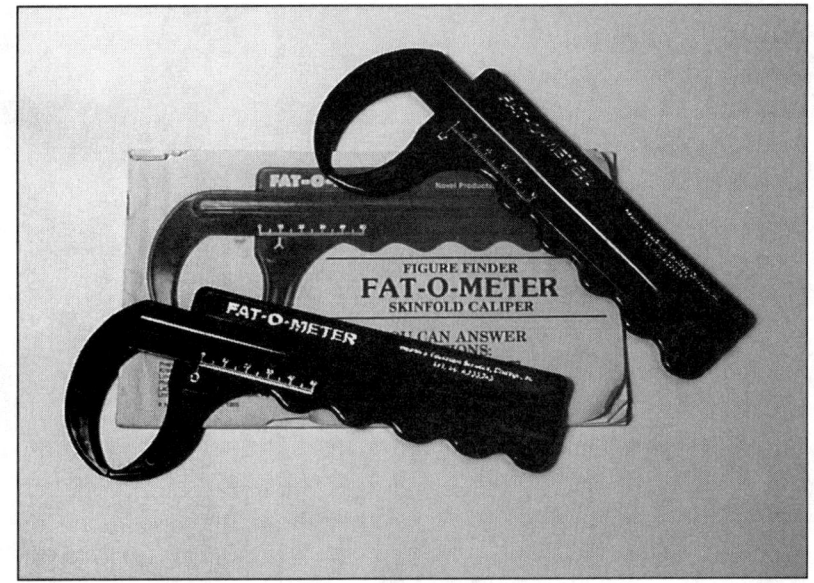

One other device that has found acceptance for percentage fat estimation is the skinfold caliper (Figure 10.10). This method involves use of a caliper to measure the thickness of pinched skin and subcutaneous fat at selected locations on the body. As 50% of the body's fat is subcutaneous, thicker folds indicate a higher percentage body fat. Several models of skinfold calipers are available, but the Fat-O-Meter by Novel Products, Inc. (P.O. Box 408, Rockton, IL 61072-0408), is the most affordable. The directions for the exercise that follows can be used for any calipers on the market, and proficiency in technique depends on following the guidelines and practice. In fact, attention to detail and practice seem to be the determinants for reliability of the method.

Determination of Percentage Body Fat

Objectives:

- To appreciate one factor used to quantify overall fitness.
- To learn to use skinfold calipers to determine body composition.
- To calculate percentage fat from the values collected at the skinfold sites.

Materials:

- Skinfold calipers and instructions.
- Calculators.

Procedure:

1. Carefully determine the three sites on the body where skinfolds will be measured. Take all measurements on the right side of the body and directly on the skin, not through clothing. These sites should be as follows:

    Males   Site 8:   Measure halfway up the upper leg from the kneecap, and pinch lengthwise on the front of the thigh (Figure 10.11a).

    Site 9:   Take a lengthwise fold of skin 1 inch to the right of the navel (Figure 10.11b).

    Site 10:   On a line from the armpit to the nipple, take a pinch halfway in between (Figure 10.11c).

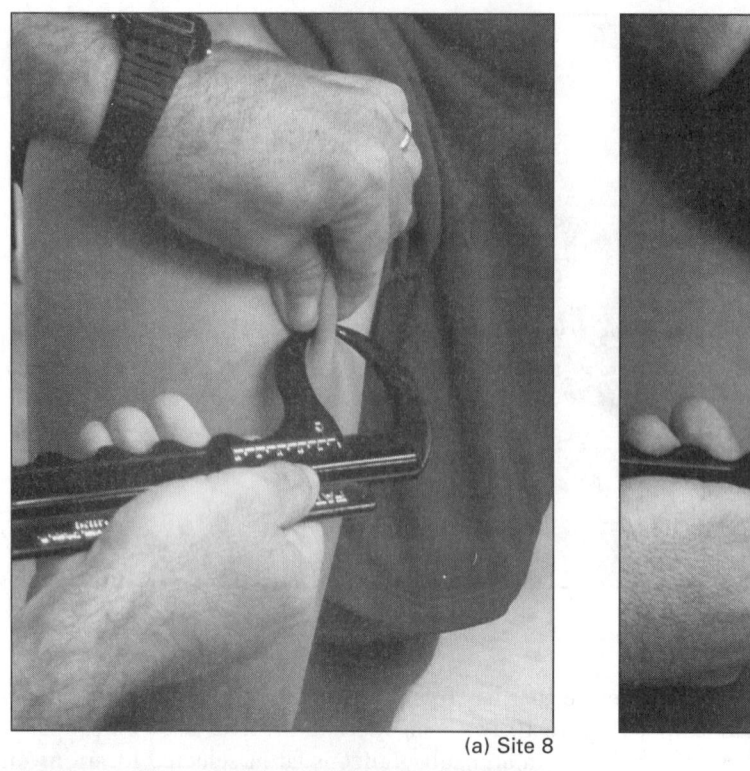
(a) Site 8

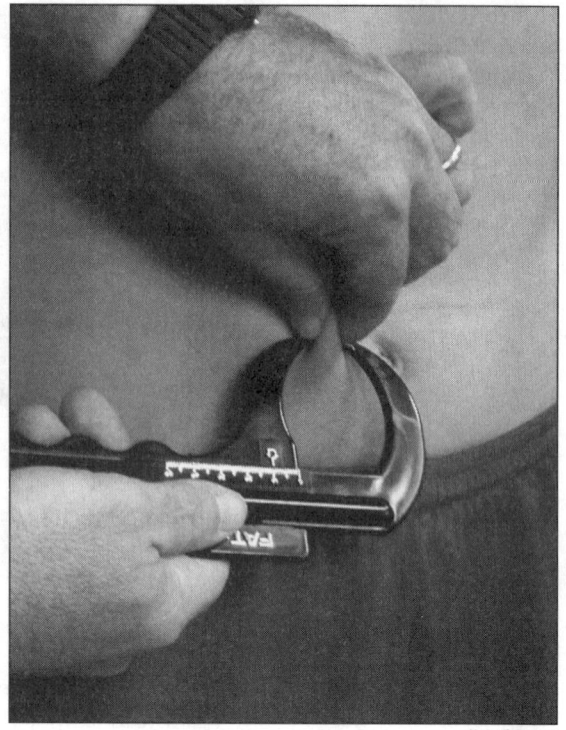
(b) Site 9

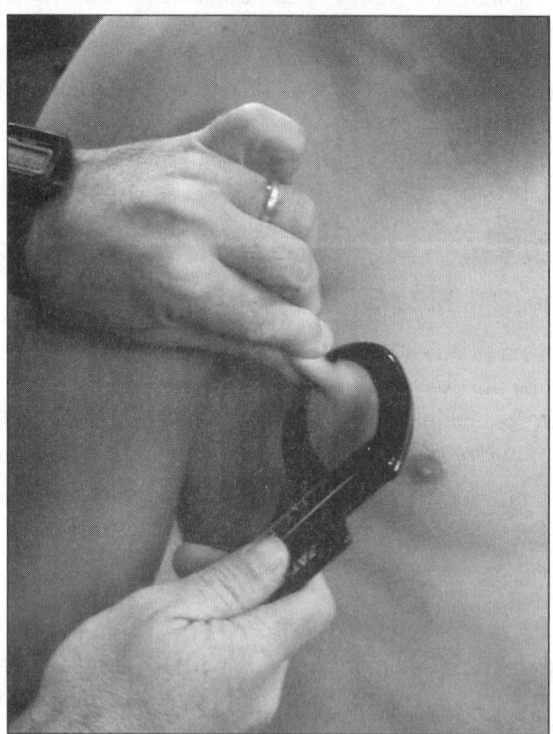
(c) Site 10

Figure 10.11
Sites for Skinfold Determination in Men.
Shown are the sites for determining skinfolds in men. (a) Site 8. (b) Site 9. (c) Site 10.

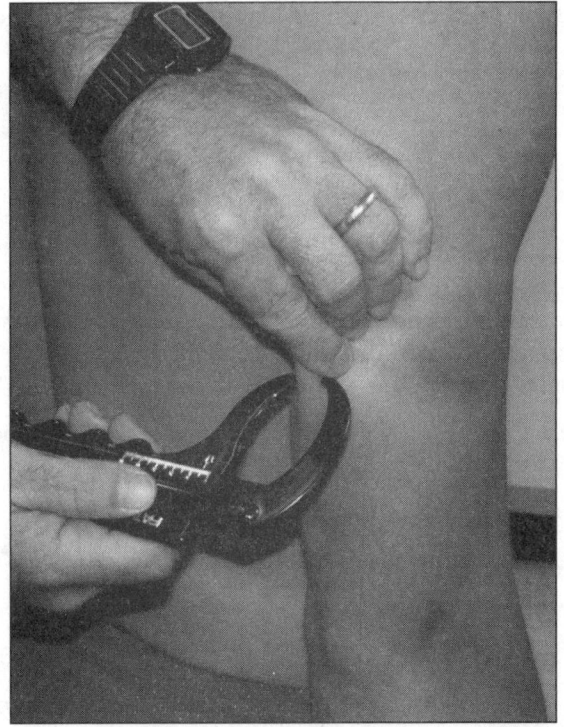

(a) Site 2

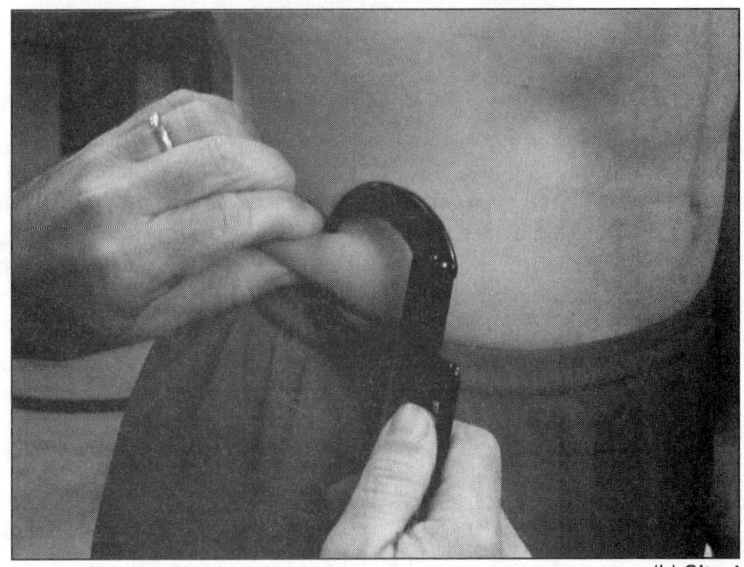

(b) Site 4

Figure 10.12
Sites for Skinfold Determination in Women.
Shown are the sites for determining skinfolds in women. (a) Site 2. (b) Site 4. (Refer to Figure 10.11a for site 8.)

Females   Site 2: Take a lengthwise pinch on the posterior of the upper arm halfway between the elbow and the shoulder (Figure 10.12a).

Site 4: Locate the ilium on the right side of the hip, and take a pinch following the natural fold 0.5 inches above the bone (Figure 10.12b).

Site 8: Measure halfway up the upper leg from the kneecap, and pinch lengthwise on the front of the thigh (Figure 10.11a).

2. Take three sets of measurements from each site, rotating through the sites in sequence. Open the calipers with one hand, and place the calipers on the pinch of skin and fat. Do not pinch the underlying muscle. Place the calipers on the fold approximately 1 inch from the fingers.

3. Select the two closest readings for each site, and average them for use in the following formulas:

Males: $X2 = \text{Site } 8 + \text{Site } 9 + \text{Site } 10$

$$\text{Density} = 1.10993800 - 0.0008267(X2) + 0.0000015(X2)^2 - 0.0002574(\text{age})$$

Females: $X3 = \text{Site } 2 + \text{Site } 4 + \text{Site } 8$

$$\text{Density} = 1.0994921 - 0.0009920(X3) + 0.0000023(X3)^2 - 0.0001392(\text{age})$$

4. After the density is determined, the density figure is used to estimate the percentage body fat. Use the following formula:

$$\% \text{ Fat} = \frac{457}{\text{Density}} - 414.2$$

Remember this is only an estimate of percentage fat in your body. There are charts and tables that give the average percentage fat for general populations, separate sexes, specific age groups, and elite athletes. The formulas presented here are designed to work for most people; more specific formulas are available in the literature. Be aware there are medical recommendations for both men and women regarding the lowest acceptable percentage body fat. For women, this is 10 to 12%; for men, it is 5 to 7%. The range for adult males is 5 to 10% for elite athletes to 20% for obesity. In adult women, elite athletes range from 10 to 15% to obesity at over 30%. The minimum figure for women is higher than that for men partly because a minimum amount of fat is required for menstruation to occur.

QUESTION: Research suggests that body composition is determined by genetics and early nutrition. Nothing can be changed regarding our genetic inclinations, but our diet can be modified. Also, overnutrition in childhood may stimulation fat-cell division. Consider the factors that complicate this issue from the standpoint of preventing overnutrition.

Observations:

Conclusion:

Journal Note Western cultures, and especially that of the United States, are very body conscious. Self-image is very important to many young people. What are some of the disorders that grow out of people's dissatisfaction with their bodies? How do you feel about your own body? How much effort do you put into making it what you want it to be? Considering what you were born with, how realistic are your expectations regarding what you would like your body to be?

Section 11

The Respiratory System

Note to the Student

One of the most obvious features of a living human being is the act of ventilation, or breathing. In old movies, to determine if the "victim" of foul play was still alive, a detective often held a mirror over the victim's mouth to see if it fogged-up, thus indicating respiration. The cardiovascular system and the respiratory system often can be thought about together, because the lungs are pointless without the heart and vessels to convey the blood that carries oxygen. Also, one of the principal functions of the circulatory system is the transport of oxygen from the lungs to the cells and tissues of the body. This section reviews the basic anatomy of the lungs as well as the passageways and tubes that supply them; it also measures some physiological parameters that relate to function.

Architecture of the Respiratory System

The nose and mouth, both of which are located on the face, are the two points of access to the respiratory system's principal organs: the lungs (Figure 11.1). Unless they are obstructed by infection or inflammation, the nose and nasal passageways generally are the preferred pathways for movement of air both into and out of the lungs.

The Nose

The term nose refers to the structure on the face between the eyes and above the superior lip. It has two openings, which are called **nostrils** or **external nares** (*nari* (L), a nostril), that allow air to pass into the nasal passageways (Figure 11.2). The nostrils actually open into an area referred to as the **vestibule** (*vestibul* (L), a porch) before entering the nasal passageways proper. The nasal passageways are lined with mucous membranes that are composed of simple columnar epithelium with numerous goblet cells and associated glands that produce a mucus of variable consistency, which moistens and cleans the air entering the body. (The importance of this cleansing function is obvious after working in any dry, dusty area, when mucus that is "blown" or sneezed from the nose will be full of dust.) Remember from your textbook that the inside of your nose is divided by the nasal septum into left and right chambers. There are some

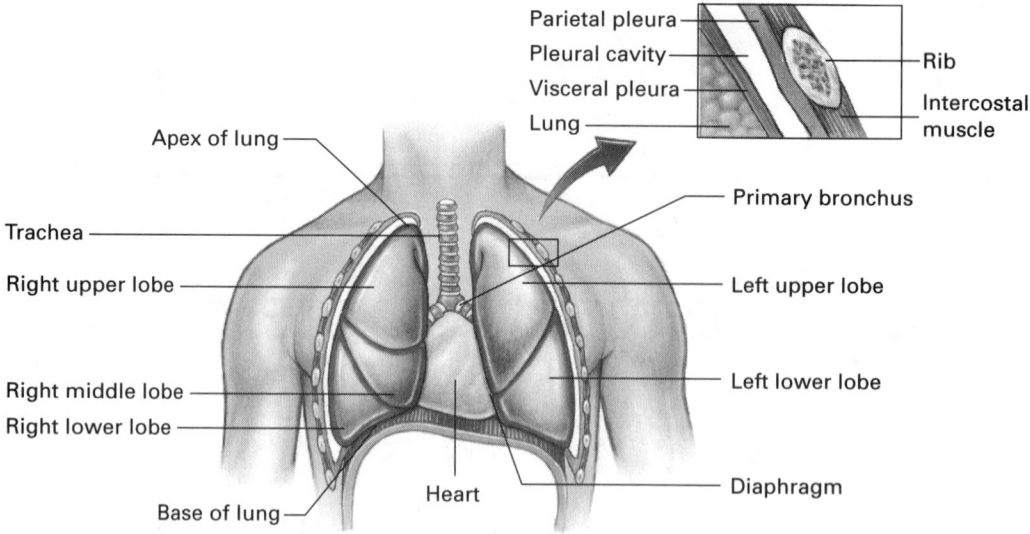

Figure 11.1
The Respiratory System.
The respiratory system consists of the lungs and their connecting tubes. The lungs deliver oxygen to the blood, which then supplies that oxygen to all the body tissues.

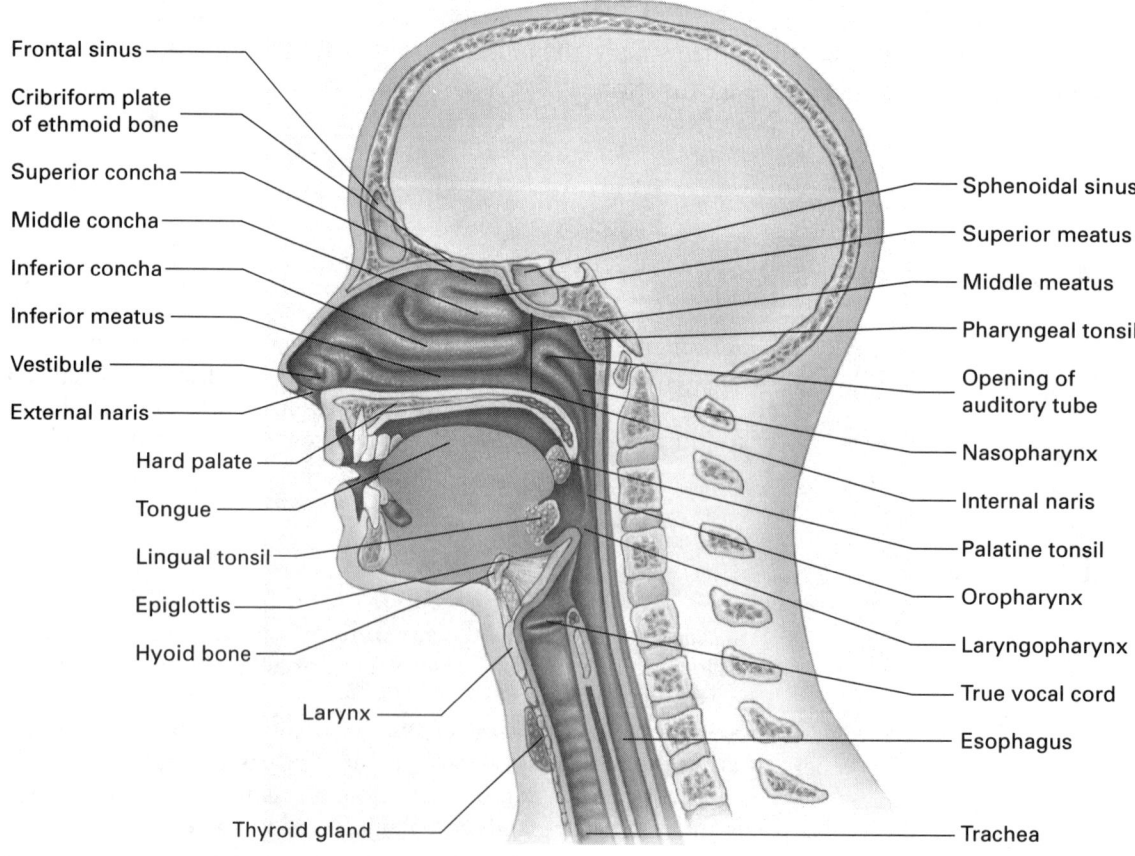

Figure 11.2
Upper Respiratory Passageways.
The nose is the opening of the respiratory system to the outside air. It connects with the nasal passageways, which open into the throat or pharynx to connect with the lungs.

additional skeletal elaborations in each chamber, the **nasal conchae,** that hang down into the passageways to increase the surface area of the mucous membranes lining the nasal passageways. These serve as baffles or walls to increase contact of inhaled air with the membranes. Finally, the exit holes of the passageways into the throat or pharynx are called the **internal nares.**

The Pharynx and Larynx

The **nasopharynx** is the part of the throat into which the internal nares open (Figure 11.2). It also is lined with mucous membranes, as are all respiratory membranes of the body except for the final oxygen-absorbing surface. The **auditory** or **Eustachian tubes** open into the nasopharynx and connect to the middle ear, and these air-filled tubes provide a pressure-equalizing connection of the middle ear with outside pressure changes. An inflamed lining of the Eustachian tubes can seal off this connection. Thus, during simple daily activity, and especially when traveling by airplane, pressure changes can be painful. The nasopharynx itself leads into the **oropharynx,** which is the section of the throat that contains the opening from the mouth. This opening is called the **fauces.** This, then, begins the part of the respiratory system that is shared with the digestive tract.

The next section of the pharynx is the **laryngopharynx,** which at its inferior end contains the opening of the esophagus into which food and drink are routed to enter the digestive tract. It also contains the glottis, or opening, through the **larynx** and into the respiratory system again. The larynx is a complicated apparatus consisting of cartilage, muscles, and connective tissue and it also is known as the voice box. The larynx contains the **vocal cords,** which actually are folds along the lateral walls of the laryngeal opening that vibrate as we exhale and produce sound. The entry of food and drink into the glottis is prevented by a flap of connective tissue called the **epiglottis.** Normally, this covers the glottis during swallowing to prevent materials being consumed from accidentally entering the respiratory system. If materials follow the wrong path into the respiratory tract, we are stimulated to cough violently to remove them from the respiratory path.

QUESTION: Often, people develop choking spells in restaurants after a piece of food becomes lodged in the "windpipe." The person becomes panicky and agitated, cannot breath, and may turn blue. This often is called a "cafe coronary" because of its outward similarities to a myocardial infarction or heart attack. With introduction of the Heimlich maneuver, however, lives have been saved by the quick action of alert bystanders or companions. How does the Heimlich maneuver work?

Observations:

Conclusion:

The Trachea and Bronchial Tree

The tube that connects the inferior aspect of the larynx with the lungs is called the **trachea** (*traches* (L), the windpipe). It is a columnar epithelium–lined tube of connective tissue that is reinforced with cartilage rings. A good analogy of the trachea is a vacuum-cleaner hose. With the lower air pressure inside, the hose must be strengthened by wires or ridges that are molded into the wall of the tube. The negative pressures generated during breathing are not great, but without the cartilage rings, the trachea would collapse from the surrounding tissue forces. The trachea branches into a left and a right **primary bronchus** (*bronchus* (G), the windpipe) to supply each lung (Figure 11.3), and each primary bronchus branches into several **secondary bronchi.** The lungs themselves are subdivided into many areas, which are called lobes, and there are as many secondary bronchi as lobes. The secondary bronchi further divide into **tertiary bronchi,** which then give rise to **terminal bronchi.** The terminal bronchi branch into **respiratory bronchioles** and, ultimately, end in the **alveolar ducts,** from which branch the **alveoli** (singular, *alveolus*) (*alveolus* (L), a cavity or pit). The alveoli are arranged in grape cluster–like groupings around the alveolar ducts and, because of their huge numbers, represent a very large surface area. There are approximately 2 m^2 of surface area on the outside of a normal-sized adult, but the total surface area of the lungs is 70 m^2.

The alveoli are composed of three types of cells. There is a type I cell, which is the squamous epithelial cell that actually makes up the alveolar wall. There also is a type II cell, which lso is known as a septal cell, that functions to produce surfactant, which is the soap-like material that reduces the surface tension of water required to moisten the internal surface of the alveolus. The third cell is the alveolar macrophage. While not technically part of the structure of the lungs, these roving white blood cells help to "police" the inner surface of the lungs to remove infectious agents, inhaled dirt, other foreign material, and injured tissue.

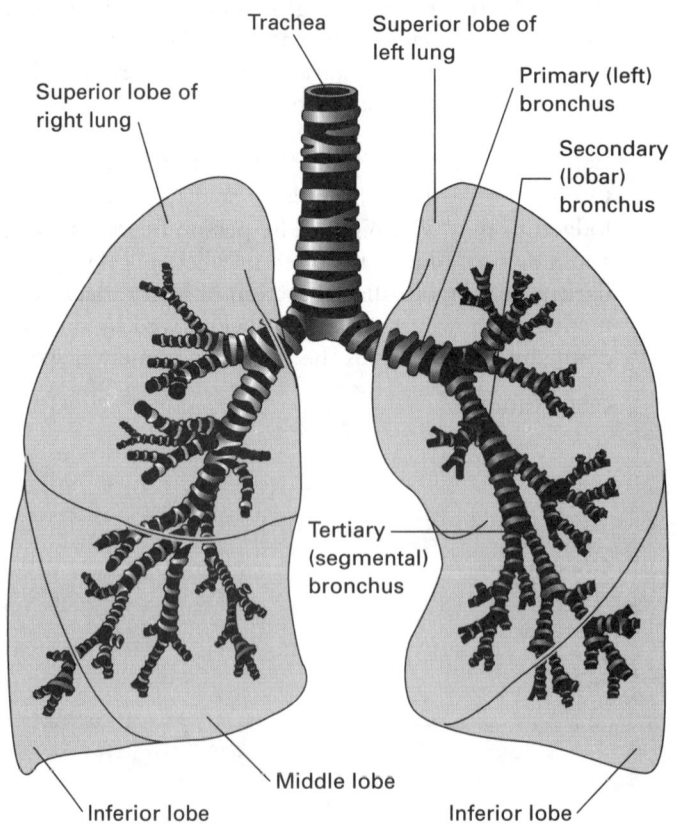

Figure 11.3
The Trachea and Bronchi.
There is a single tube, the trachea, connecting the laryngopharynx with the lungs. It branches into a left and right bronchus, which open into the respective lung. Secondary bronchi arise from the primary bronchi and connect with each lobe of the lung (three for the right lung and two for the left).

QUESTION: Because of accidental or occupational exposure, persons often breath in fine dust particles such as silica and asbestos. Macrophages can eat and digest most organic materials, but these particulates are undigestible. What happens to a macrophage that eats something it cannot digest? What happens to the lung tissue in someone who has too high a contamination with this type of dust?

Observations:

Conclusion:

Respiratory Muscles and Breathing Control

The process of breathing is controlled by several centers in the brain stem. Many questions still remain about the exact mechanism of respiratory control, but it certainly involves the **medullary rhythmicity center** in the medulla and the **apneustic** and **pneumotaxic centers** (*pneumo* (G), the lungs; *pneusis* (G), breathing) in the pons. Regardless of the mechanism, the muscles involved in the process of breathing are the diaphragm, the intercostals, and secondarily, the abdominal and shoulder muscles (Figure 11.4). Most of the work involved with resting respiration is done by the diaphragm; however, during increased activity, other muscles become active participants. Resting expiration primarily is by relaxation of the breathing muscles.

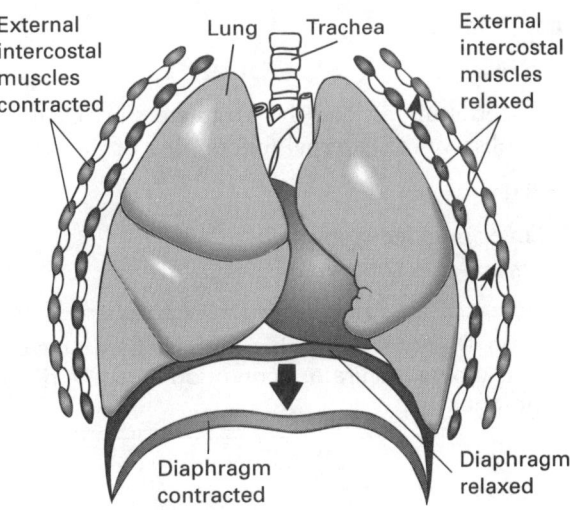

Figure 11.4
The Muscles of Breathing.
The diaphragm is the principal respiratory or breathing muscle. The intercostals, abdominals, and shoulder muscles also participate in normal breathing movements.

QUESTION: All of us have exerted ourselves playing a favorite sport to the point of having to stop and "catch our breath." What muscles are actively involved in the process of breathing when we exercise?

Observations:

Conclusion:

The body monitors levels of carbon dioxide to determine the efficiency of breathing. If the level is increasing, respiratory control centers respond by increasing the rate and depth of breathing. If the level is decreasing, respiratory effort then is more than adequate to meet the body's needs, and the rate and depth of breathing consequently are slowed. Under pathological conditions associated with disease, however, the level of oxygen becomes the drive for breathing.

Breath Holding

We hold our breath when we dive into a swimming pool, to stop the hiccoughs, and, at least when we were young, even to get our way: What determines how long we can hold our breath? As mentioned, the carbon dioxide and oxygen concentrations in the blood influence the breathing rate and depth. Therefore, how long we can hold our breath also depends on the concentrations of these gases in the blood.

Breath Holding

Objectives:

- To better understand the parameters involved in the regulation of breathing.
- To observe the external functioning of the respiratory system.

Materials:

- Timer or clock/stopwatch.
- Oxygen tank (optional).

Caution: Do not attempt this exercise if you have ever had difficulty breathing, have a known respiratory condition, become dizzy or nauseous easily, or may be pregnant.

Procedure:

1. The person holding his or her breath should not look at the timing device during this exercise. Therefore, there will be a timer and a subject. Caution should be exercised to avoid overdoing breath holding. Do not try to show off. Timers should watch subjects carefully to prevent them from passing out.

2. The subject should take one normal breath and hold it. The timer should note the "start" time. Watch the subject, and record the time at which the subject can no longer hold his or her breath.

   Time that the breath can be held after normal breath: _____

3. Instruct the subject to take several deep breaths and then hold one. Time the period of breath holding.

   Time that the breath can be held after several deep breaths: _____

4. If oxygen is available in the lab, take several deep breaths of oxygen (the same number as in step 3) and then hold one. Record the time for breath holding.

   Caution: Do not take more than a few (2 or 3) breaths of pure oxygen for this test.

   Time that breath can be held after several breaths of oxygen: _____

QUESTION: Based on the involvement of carbon dioxide and oxygen in control of breathing, why can the breath be held for different periods of time following different experimental set-ups.

Observations:

Conclusion:

Journal Note As a child you may have tried to hold your breath for a long time so you could swim all the way across a pool underwater. To do this you probably took several large breaths before diving under the water. Can you explain how these several large breaths helped you to hold your breath for a longer time? Are there any hazards associated with holding your breath too long?

Respiratory Volumes

With each breath, a certain volume of air is moved into or out of the lungs. This air is referred to as the **tidal volume (TV)** (Figure 11.5). A typical value for tidal volume is 500 ml. With exercise, tidal volume increases, thus suggesting there is additional airspace in the lungs to be used when demand requires. This additional volume is known as the **inspiratory reserve volume (IRV)** and the **expiratory reserve volume (ERV)**. Normally, these are 3000 and 1000 ml, respectively. This means there is an additional 4000 ml of volume in the lungs that can be used to meet functional demands when needed. The reserve volumes are determined by measuring the maximum inhalation or exhalation that can be achieved after a normal inhalation or exhalation. The sum volume of TV + IRV + ERV equals the **vital capacity (VC)**. This, therefore, is the volume that is available to maintain the metabolic machinery of the body. These volumes all can be measured with **spirometry** (*spiro* (L), breathe; *metry*

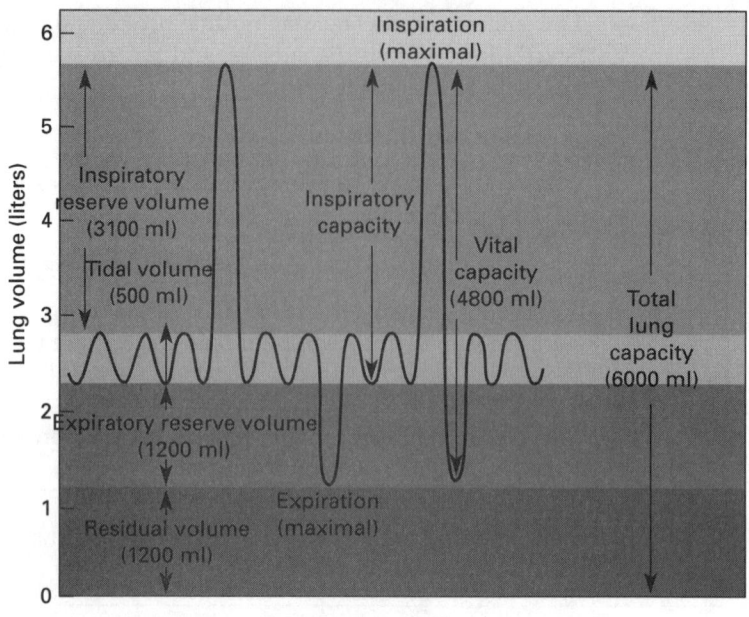

Figure 11.5
Respiratory Volumes.
This schematic shows the various parameters that can be measured with a recording respirometer. Note that the inspiratory reserve volume, tidal volume, expiratory reserve volume, vital capacity, and breathing rate are readily determined from such a recording.

(G), measurement) equipment. The vital capacity does not represent the entire volume in the lungs, and the lungs are not collapsed even as we expire our maximum expiration. The volume remaining in the lungs after a maximum expiration is called the **residual volume (RV)** (normal value, 1500 ml). This is the volume of gas that fills the parts of the lungs—the alveoli, bronchial tree, and trachea—that still contain gas even when the lungs are "empty." Residual volume cannot be determined by spirometry; therefore, neither can the **total lung capacity (TLC)** (normal value, 6000 ml) be determined this way. Total lung capacity, however, is the total volume of the lungs, meaning the vital capacity and the residual volume (VC + RV = TLC).

Spirometry to Determine Respiratory Volumes

Respiratory volumes are useful clinically to determine if disease is compromising the respiratory system. The device used to determine these volumes is called a spirometer or a respirometer. There are very inexpensive models that only allow a few, basic measurements as well as very expensive models like those used in clinical or hospital settings (Figure 11.6). The number of measurements that you can do will depend on the specific device in your lab.

Respiratory Volumes

Objectives:
- To understand the basic methodology for evaluating characteristics and functioning of the respiratory system.
- To understand the various respiratory volumes.

Materials:
- Spirometer or respirometer.
- Alcohol pads.
- Calculator (optional).
- Medical oxygen (optional).

Figure 11.6
Different Types of Respirometers.
There are several brands and models of recording respirometers, which are capable of a variety of clinically important respiratory measurements. Shown is the Student Wet-Spirometer by Phipps & Bird, Inc. (1519 Summit Avenue, Richmond, VA 23230) and the Collins 9-Liter Respirometer by Warren E. Collins, Inc. (220 Wood Road, Braintree, MA 02184-2404).

Procedure:

1. Study the instrument, and learn how to use the various valves and controls. Select a subject and an observer (some persons have difficulty breathing whenever there is any added resistance to the breathing apparatus), and fill the instrument with air or oxygen. Obtain a clean mouthpiece for use by the subject. You may want to clean the mouthpiece with an alcohol pad. Nose clamps are recommended as well, because the subject likely will allow air to leak from the nose while breathing through the mouthpiece. (There also may be significant salivation because of the mouthpiece in the mouth, so be prepared.)

Caution: If the test will be of short duration, room air can be used. If the device is a recording device and, therefore, will allow more sophisticated measurements, oxygen will provide a margin of safety over a longer period of testing. Room air, however, obviously contains enough oxygen to sustain life. As mentioned, carbon dioxide is important in the regulation of breathing. If the test will last longer than one minute, the device should have a way to absorb carbon dioxide from the test chamber to prevent it from influencing the rate and depth of respiration.

---

2. Decide how long the test will last, but be prepared to stop if the subject experiences any difficulty. Attempt to measure the following parameters, which are readily determined with any recording respirometer (Figure 11.7):

Parameter	Normal value
Tidal volume	500 ml
Inspiratory reserve volume	3100 ml
Expiratory reserve volume	1200 ml
Inspiratory capacity	36500 ml
Vital capacity	4800 ml
Breathing rate	12–16 breaths/min

Figure 11.7
Respirometer Recording.
The different respiratory volumes are indicated on this actual recording from a respirometer.

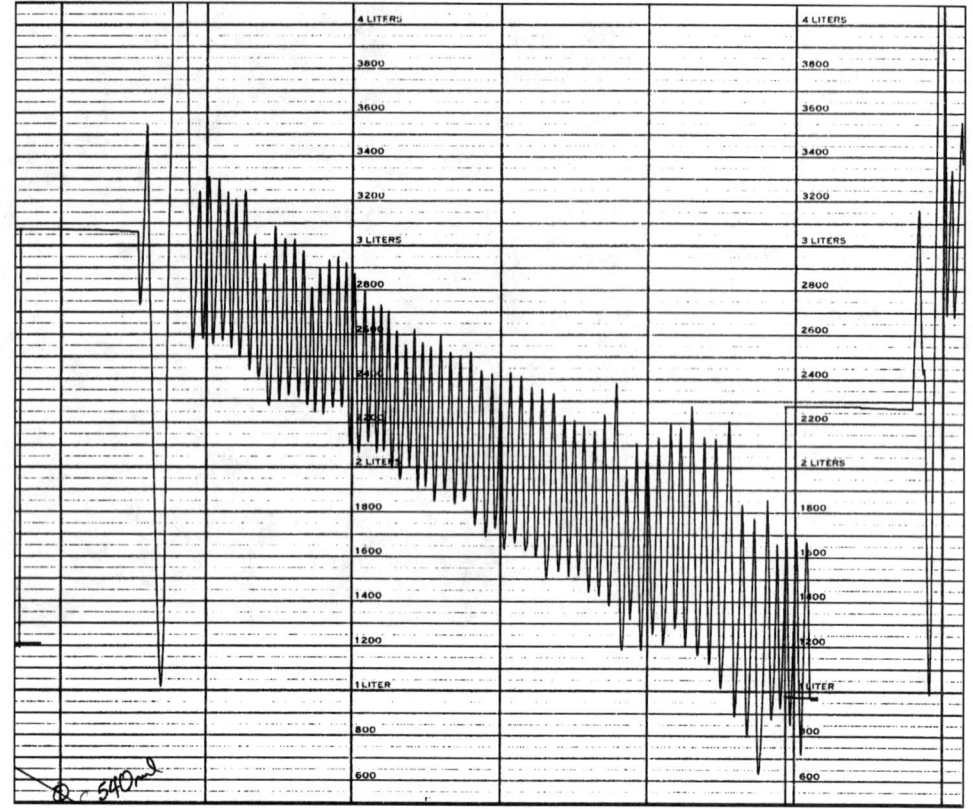

3. Complete the calculations required to determine the values listed. The recording paper of the respirometer will be calibrated horizontally with the paper speed of the motor and probably will be marked in minutes at the slowest speed. The vertical marks on the paper will be calibrated in units of volume so that the listed parameters can be determined.

QUESTION: Emphysema and chronic bronchitis both are conditions that greatly affect the respiratory function. Emphysema results in fusion of alveolar spaces into larger openings, and this reduces the surface area but increases the volume of the affected areas of lung. Chronic bronchitis is an inflammatory disease of the airways. Which do you think is most likely to affect the respiratory volumes?

Observations:

Conclusion:

Journal Note Do you, a close friend, or a relative smoke or use products containing tobacco? What health problems are linked to such use? Has media coverage of the health problems associated with tobacco products had an effect on you? Do you see as many, more, or fewer people using such products as in the past? Has the age distribution of these users changed? Are more women using these products than in the past?

Determining the Metabolic Rate

The rate of energy utilization is an important consideration for survival. Efficient use of the resources available to an organism should help in its competition with others. Practically speaking, for humans living in developed parts of the world, energy utilization has more to do with whether we gain or lose weight rather than whether we survive. Dietary and nutritional considerations, however, figure prominently in the final recommendations of numerous studies regarding factors that affect how long we live and what diseases we may develop. To directly measure the energy metabolism of an organism would require detecting the amount of heat energy coming out of the organism, but because air is so temperature unstable, it is virtually impossible to accurately determine heat energy coming out of a human body. Therefore, energy metabolism is not routinely measured in humans. Instead, the metabolic rate of humans is determined indirectly by measuring oxygen consumption and then calculating the metabolic rate using a conversion factor.

Practical Considerations in Metabolic Measurements

Even indirect determinations of the metabolic rate in humans can be difficult, or at least expensive. Human-sized, accurate respirometers can easily cost several thousand dollars. Small, animal devices are very inexpensive, however, and allow demonstration of the same basic techniques as used for humans. Such a unit is available from Phipps & Bird, Inc. (Figure 11.8).

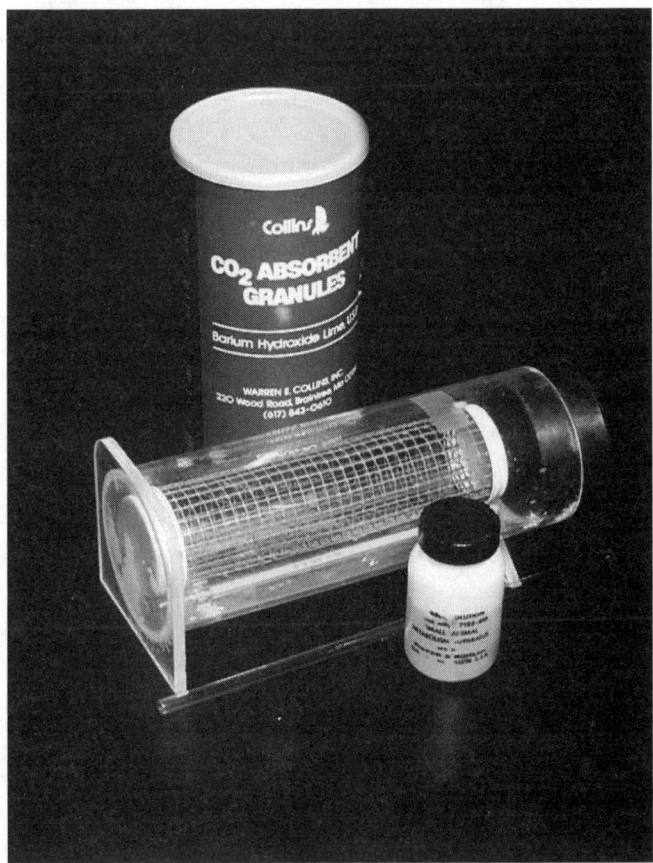

Figure 11.8
Small Animal Metabolism Chamber.
This chamber manufactured by Phipps & Bird, Inc. (1519 Summit Avenue, Richmond, VA 23230), is designed to allow metabolic measurements with any terrestrial animal that can fit comfortably into the cage of the chamber.

Metabolic Measurements in Small Animals

Objectives:
- To learn the techniques involved in metabolic measurement.
- To become familiar with handling small mammals.
- To calculate the metabolic rate and relate it to the daily energy requirements of an animal.

Materials:
- Small rodents, such as young rats, full-grown mice, or gerbils.
- Small Animal Metabolism Apparatus (Phipps & Bird, Inc.), including:
  Metabolism chamber
  Animal cage container with removable ends
  Large rubber stopper
  Calibrated tube
  Thermometer
  Bubble solution
  or
  Small glass or plastic chamber with separate animal enclosure that can be closed with a stopper, plus other items mentioned.
- Soda lime (carbon dioxide absorbent).
- Balance to weigh animal.

Procedure:

1. Prepare the metabolism chamber by placing a small amount (50 ml) of soda lime in the bottom. This will absorb the carbon dioxide produced by the animal in the chamber.

   Caution: Soda lime is hazardous. Do not allow this material to contact the animal or yourself.

QUESTION: Carbon dioxide is present at very low levels in the air and is not dangerous at these concentrations. At high concentrations, however, carbon dioxide can be fatal. Ignoring these two considerations, why is it necessary to eliminate the carbon dioxide from the metabolism chamber?

Observations:

Conclusion:

2. The weight of the animal is necessary for the calculations. The animal may be weighed before or after the measurements. If the animal has not been handled consistently, weighing after testing will prevent weighing-associated disturbance from affecting the readings.

   Weight of animal _____

3. Place the animal in the cage. Place the cage in the chamber, and close the chamber with the rubber stopper. Cover the chamber with a paper-towel tent to avoid activity in the room from disturbing the animal. Allow the chamber and the animal to temperature equilibrate for approximately 15 minutes.

4. Wet the inside of the tube with water or the bubble solution. Then, suck a single bubble from the frothy surface of the solution into the tube.

5. Insert the tube into the hole in the stopper. This seals the chamber from the outside air and allows the oxygen consumption to be measured.

6. As soon as the tube is inserted, time how long it takes for the bubble to move a certain distance in the calibrated tube. After a measurement, remove and reinsert the tube after reversing it 180o, then repeat the measurement. Record at least three measurements.

Measurement	Volume Consumed	Time
1	_____	_____
2	_____	_____
3	_____	_____
Total	_____	_____

7. Remove the animal from the container as soon as the measurements are completed. Do not pull forcefully on the end of any rodent's tail, however, because it can break off. Removing both ends of the cage and gently coaxing should be sufficient to remove the animal. These animals like to enter small spaces, so surrounding the end of the tube with a towel or rag should encourage it to leave the tube.

8. Determine the volume of oxygen consumed per unit time. To do so, divide the total volume (in milliliters) by the total time (in minutes).

   Oxygen consumed = Volume of oxygen / Time

   Oxygen consumed (ml/min) = _____

9. Because temperature variation and changes in barometric pressure can affect your results, the results often are corrected to the standard temperature and pressure (0°C and 760 mm Hg). This is not necessary for our purposes.

10. To convert oxygen consumption to the metabolic rate requires a conversion factor. This is 4.8 calories per 1 ml of oxygen. Therefore, to convert, simply multiply the oxygen consumption by the conversion factor.

    (Oxygen consumption) × (4.8 cal/ml oxygen) = _____ cal/min

11. Warm-blooded animals lose most of their heat through their skin or surface area. Therefore, a correction normally performed with this value just calculated is to divide it by the total surface area of the animal. This would change the units to cal/cm2/min, which are the units in which metabolic rate is reported. Because this requires an additional chart, however, we will not carry out this calculation.

QUESTION: An important relationship in biology is that as an animal grows larger, its surface area increases by the square, but its volume increases by the cube. Therefore, larger animals have more volume per surface area than smaller animals. Based on this observation, should larger animals have higher or lower metabolic rates than smaller animals?

Observations:

Conclusion:

QUESTION: Conversion factors must be appropriate for the factors under study. The conversion factor for the metabolic rate above is 4.8 calories per 1 ml of oxygen. What should the conversion factor be for larger animals such as humans? (Hint: Quantity doesn't change but units should change to fit a larger animal.)

Observations:

Conclusion:

Oxygen-Carrying Capacity of Blood

One of the major functions of blood is to carry oxygen. Red blood cells (RBCs) essentially are cell membrane–bound structures that contain millions of hemoglobin molecules. It actually is the hemoglobin that binds to and carries oxygen to the tissues of the body. Because hemoglobin has a red color and binds oxygen, it is known as a respiratory pigment. The adequacy of blood to carry oxygen for the body is a function of the number of RBCs and the concentration of hemoglobin in the blood. The ability of blood to perform its respiratory function can be measured because the hematocrit, RBC count, and hemoglobin concentration all can be readily measured.

Hematocrit is defined as the percentage of whole blood that consists of formed elements, which essentially refers to the cells that are in the blood. The term cells is not used, however, because the platelets, which are cell fragments and not whole cells, are included in the hematocrit.

Total Red-Blood-Cell Count

The total number of RBCs in the blood is an important determinant of its oxygen-transporting potential. As explained later, the amount of hemoglobin in the cells also is important. When counting the RBCs, the major concern is that each microliter of blood has so many of them. Therefore, the only way to count RBCs is to take a small blood sample and then dilute it significantly. The same device used for counting the white blood cells (WBCs), the hemocytometer, also can be used to count RBCs.

Hemoglobin Concentration

Hemoglobin is the actual chemical in the blood that carries oxygen to the cells. Hemoglobin is not free in the bloodstream; it is contained in the RBCs. Hemoglobin is a very complicated molecule, and it is capable of binding oxygen as well as responding to physiological demands in its loading and unloading of the oxygen.

QUESTION: Hemolysis refers to the breaking down of RBCs. Sometimes, the plasma or serum collected has a slight pink to red coloration. What action often used in collection of blood from a finger might lead to hemolysis?

Observations:

Conclusion:

Suggested Activity: Complete Blood Count/Lab Work-Up

Note: This activity has implications for the discussion of the circulatory system in Section 9. It is presented here, however, because some of the most common items routinely analyzed in these tests deal with the RBCs and their impact on the respiratory system.

Your instructor might arrange for an activity dealing with the many different items that can be measured in blood. This section already has described important parameters often studied using blood tests. A complete blood count (CBC) often contains two large groups of measurements: hematology, and chemistry. Hematology covers parameters such as total RBC and WBC counts, hemoglobin concentration, hematocrit, platelet count, and differential WBC counts. Depending on the particular medical problem that is suspected, chemistry can be quite extensive. It may include measuring levels of billirubin, calcium, sodium, potassium, total cholesterol, low- and high-density lipoproteins, iron, triglycerides, and many more. Each lab typically has a sheet that lists the normal values for these parameters. These normal values will vary a little from lab to lab depending on the particular technique used.

If possible, a volunteer from the class will undergo a CBC and blood chemistry work-up at a local lab. The results will be obtained and discussed in lab. A technician also may be invited to point out any deviations from the normal values and their significance.

Journal Note Have you ever had a CBC and blood chemistry performed as part of a physician visit for a medical problem or pre-employment physical? Were the results normal? Were any specific values out of the normal range and, therefore, indicative of a homeostatic imbalance? How much did the test cost? Some aspects of chemistry tests are expensive, so they are not performed on a routine basis. Have you ever had any expensive tests? If so, for what medical problems? Did the medical personnel fully explain the rationale for the test? Did you ask questions? Do you think that people turn over too much responsibility for their own health to their physicians?

Section | 12

The Urinary System

Note to the Student

The urinary system and its principal organ, the kidney, represent one of the more fascinating areas of anatomy and physiology. The overall structure of the urinary system is simple (Figure 12.1), but the microanatomy of the kidney is unique. Along with being the sanitizing filters of the body, the kidneys also control several other important physiological aspects of the body's operation. For example, they are involved in regulation of blood pressure and stimulation of RBC synthesis. Each kidney contains approximately 1 million **nephrons** (*nephrus* (G), the kidney), which are considered to be the structural and functional units of the kidneys. The term nephron describes a set of tubes and blood vessels that accomplish the work of the kidney; this means an understanding of the structure and function of nephrons will show how the kidney is organized and works. The urinary system also contains tubes that drain fluid from the kidneys and a bladder to store that fluid before it is eliminated from the body. The system differs a bit between the genders as well, in that some of the tubes in men serve both a urinary and a reproductive function.

The Kidneys

The urinary system contains two kidneys, which are located along the posterior abdominal wall on each side of the body, approximately at the level of the most inferior ribs. The right kidney is slightly lower because of the mass of the liver, which is positioned just above it (Figure 12.1). Because the kidneys are not completely surrounded by the peritoneum that lines the abdominal cavity and its organs, their position is referred to as being **retroperitoneal** (*retro* (L), back, behind). They are bean-shaped organs, approximately 5 inches long, and weigh 150 g each (Figure 12.2). The medial margin of the each kidney contains a depressed or indented area called the **hilum** (*hilum* (L), a little thing), which is where the **renal arteries, renal veins,** and **ureter** (*ren* (L), a kidney) either enter or exit the kidney. The outermost covering of the kidneys, or **renal capsule,** is composed of fibrous connective tissue.

The kidneys are unique in that some of their structural features are apparent to the naked eye. If a kidney, either human or one from another mammal, is sectioned along the frontal plane, two areas are apparent in the kidney tissue proper (Figure

Figure 12.1
The Urinary System.
The urinary system contains a pair of kidneys and ureters, a single bladder, and the urethra.

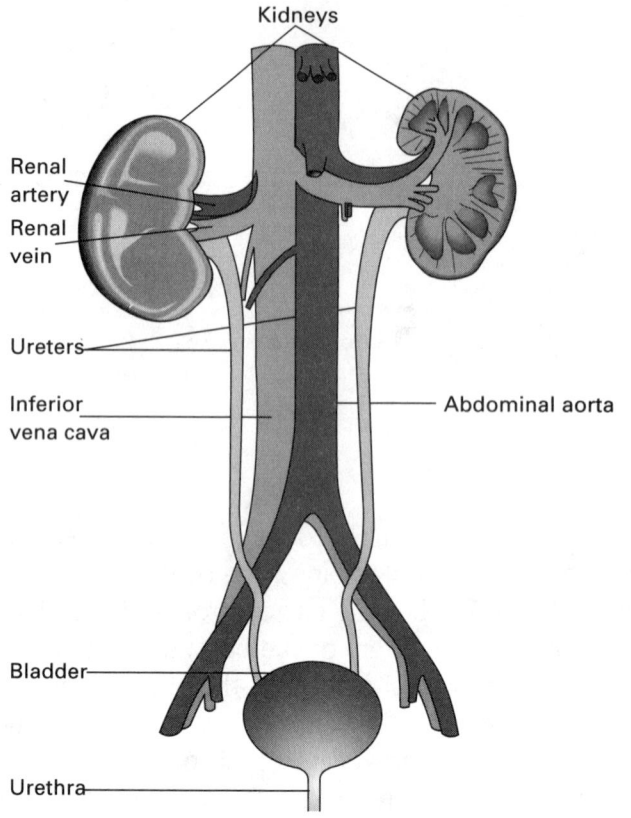

Figure 12.2
Frontal Section of the Kidney.
Note that areas labeled inside the kidney show differences in their appearance based on the arrangements of the nephron components.

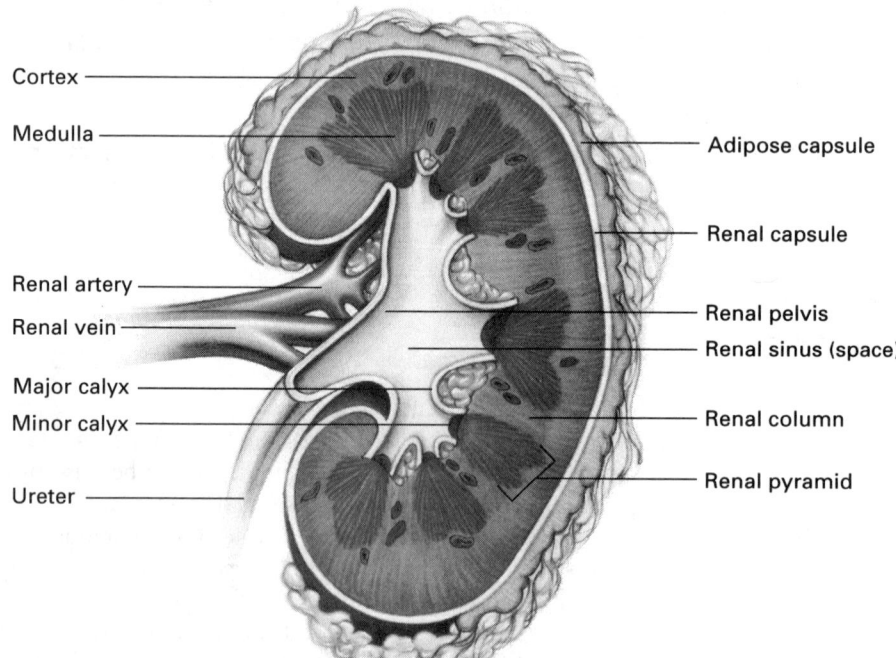

202　SECTION 12　The Urinary System

12.2). There is an outer region, the **renal cortex** (*cortex* (L), the bark), and an inner region, the **renal medulla** (*medulla* (L), marrow). The medullary region contains triangular-shaped structures called **renal pyramids,** with their peaks or apexes pointing toward the hilum. The pyramids have a striped appearance because of the internal arrangement of certain parts of the nephrons. Areas of the cortex tissue extend between the pyramids and are referred to as **enal columns.** The cortex has a granular appearance, again because of the arrangement of similar parts of the nephrons in certain areas. The fluid, or **urine,** the kidneys produce drains from each renal pyramid into a cup- or funnel-shaped structure called a **calyx** (plural, *calyces*) (*calyx* (G), cup-like animal structure). Several **minor calyces** drain into a **major calyx,** and the major calyx drains into the **pelvis** (*pelv* (L), a basin), which then narrows and becomes the ureter. The ureters from both kidneys then drain into the **urinary bladder,** which contains several layers of smooth muscle in its wall. The urinary baldder holds approximately 500 ml, but it can hold double that when needed because of its highly expandable lining.

QUESTION: One characteristic change in women during pregnancy is the need to urinate more frequently. What causes this change in bladder function? Is there some change in the rate of urine formation, or is there an anatomical explanation?

Observations:

Conclusion:

The bladder and the ureters are lined with transitional epithelium, which flattens as it stretches to maintain the seal of the bladder wall (Figure 12.3). The wall of the bladder also has folds, or **rugae,** that allow for expansion when necessary. The **urethra** is the tube that drains urine to the outside of the body, and two valves, the **internal urethral sphincter** and the **external urethral sphincter** (*sphinct* (G), bind tight, squeeze), keep the bladder closed until the person is ready to urinate.

QUESTION: Other openings from the body to the outside also have valves to control what passes from the body. For the two valves just described, which contains smooth muscle, and which contains skeletal muscle? Which sphincter is under voluntary control, and which is under involuntary control?

Observations:

Conclusion:

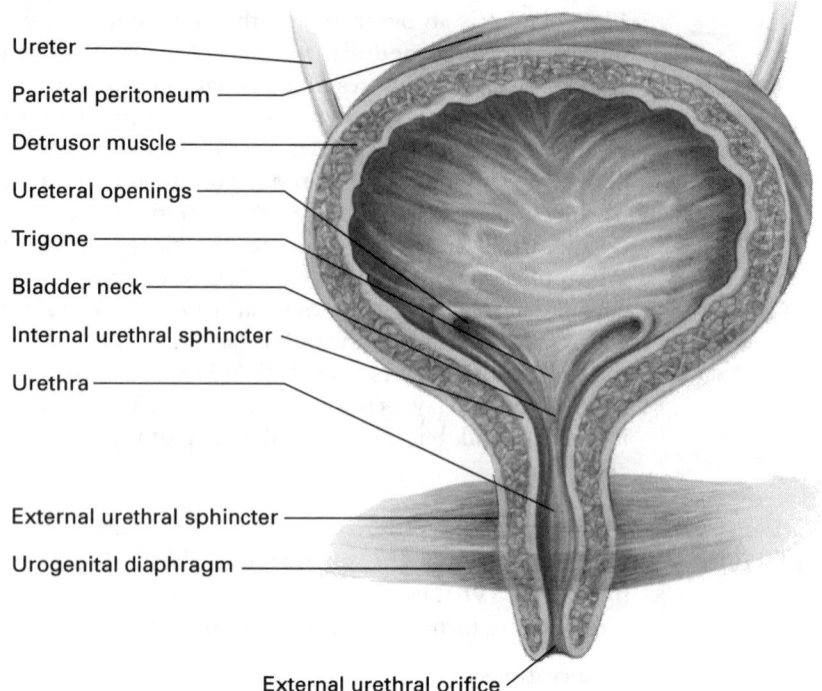

Figure 12.3
Section of the Bladder.
The wall of the bladder contains smooth muscle, and the lumen is lined with transitional epithelium, which stretches as the bladder fills.

The urethra is shorter in women than in men. This is because of the sexual function of the urethra in men. The female urethra is approximately 3 to 4 cm in length; the male urethra, because it courses through the penis, is approximately 20 cm in length.

QUESTION: The female urethra is much shorter than the male urethra. Which gender do you think is most likely to suffer from urinary tract infections? Do you have any anecdotal or personal experience to influence your conclusion?

Observations:

Conclusion:

Structure of the Nephron

As mentioned, understanding the structure of the nephron is the first step in understanding how the kidneys filter and clean the blood. The nephron is a set of tubes and blood vessels whose arrangement allows filtration, reabsorption, secretion, and concentration of the filtrate to produce the urine (Figure 12.4). These blood vessels branch inside the kidneys to supply all its parts, and they give rise in the renal cortex

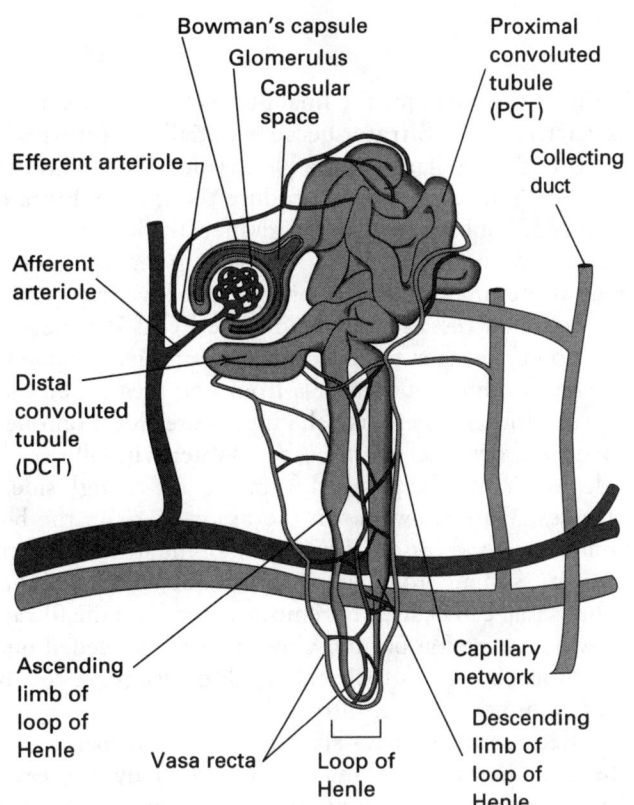

Figure 12.4
Nephron.
The nephron is the structural and functional unit of the kidney. The nephron shown has been opened for study. The 1 million nephrons present in the kidney are intertwined with each other.

to vessels called **afferent arterioles.** The afferent arteriole of each nephron branches in turn to give rise to a tuft of blood vessels called the **glomerulus** (plural, *glomeruli*) (*glomer* (L), a ball of yarn). These blood vessels actually are open capillaries, and they allow filtration, the first step in urine formation, to occur. The driving force behind filtration is blood pressure and because of the short, large renal arteries, blood reaching the afferent arteriole is under higher pressure than in the capillaries in the rest of the body. Blood drains from the glomerulus into the **efferent arteriole,** and cupping the glomerulus is the **Bowman's capsule,** which is a funnel-shaped structure that collects the filtrate formed at the glomerulus.

The filtrate then flows into the **proximal convoluted tubule (PCT).** The PCT is closer to the beginning of the nephron structure, so it is the "proximal" structure. Reabsorption occurs in the PCT. Filtration is a passive process, in that virtually everything that is small enough to be filtered is in fact filtered. This includes many important and biologically valuable substances, such as amino acids, sugars, and all ions, including Na^+, K^+, Ca^{+2}, and Cl^-. Reabsorption allows the body, at the expense of energy, to recover many important materials so that the body will not lose them. The PCT drains into the **loop of Henle,** which consists of a **descending** and an **ascending limb.** The function of the loop of Henle is described later, as part of the last step in urine formation, but the loop connects with the **distal convoluted tubule (DCT).** Both convoluted tubules are referred to as "convoluted," because they are not straight. Rather, they are curved and contorted throughout the kidney tissue. The DCT drains the urine into a **collecting duct,** which empties into a minor calyx, and then into a major calyx, the pelvis, the ureter, the bladder, and finally, out of the body via the urethra.

Urine Formation

Urine is the name for the final product that flows from the kidney. As urine forms, it is referred to as **filtrate,** because initially, it forms by the filtering of fluid through holes in the capillaries of the glomerulus. Thus, filtration depends on adequate blood pressure and blood flow to produce the filtrate. Filtrate is formed at the phenomenal rate of 120 ml/min for all the glomeruli added together. On a daily basis, this would be 180 L/day. Obviously, another step in the process reduces this volume. This step is reabsorption. Through the process of active transport, most filtered materials return to the blood vessels surrounding the PCT. Whenever solute is "pumped" from one location to another, the concentration gradient changes; remember that the difference in concentration of materials from one area to another is referred to as a gradient. These differences are like the high-water level on one side of a dam compared with the low-water level on the other. Water will fall over the dam from the high to low side, but it must be pumped from the low to high side. The water concentration also changes. Water flows down the gradient, or in the body, out of the PCT into surrounding blood vessels. Thus, 99% of the filtrate eventually is taken back into the circulatory system from where it was just filtered. The loop of Henle concentrates solutes that can be used to remove water from the filtrate, which then enters the DCT, in which secretion occurs. Unwanted or unneeded quantities of normally occurring substances as well as toxic chemicals or drugs are removed from the bloodstream and placed into the flow of filtrate.

Because the kidneys are the body's principal organ for water and salt regulation, the amount of water and salt in the body a greatly influences kidney function. Excessive water consumption typically results in increased output of water; this is known as **diuresis,** or water loss. If there is excessive salt consumption, the kidneys eventually will rid the body of the excess salt. Water is the passive player in all these processes. As a result, if there is excessive salt in the body, there will be water retention until the body gets rid of the extra salt. Thus, the body's content and intake of water or salt will affect the volume and concentration of urine.

Effects of Fluid Intake and Exercise on Urine Output

Objectives:

- To gain experience working with bodily fluids other than blood and saliva.
- To observe some clinically important characteristics of urine.
- To identify factors involved in the regulation of urine production.

Materials:

- Latex gloves.
- 1 l of distilled water (ice, optional).
- Urine collection cups.
- Urinometer cylinder.
- Hydrometer.
- Graph paper for plotting results.

Procedure:

Caution: Urine is a bodily fluid; therefore, exercise care in its handling and disposal. Wear gloves when handling urine samples, avoid spilling urine, and wipe up any spills that do occur. Do not dispose of large quantities of urine in the sink. Clean glassware thoroughly, or place it in the special wash containers provided.

1. Half the class will "volunteer" as subjects; the other half will analyze the data. At the beginning of the lab, subjects will be instructed to empty their bladders.

2. Subjects will consume 1 l of distilled water. Ice added to the water may make consuming this large volume more pleasant. Record the time that the water is consumed. Each subject should try to consume the 1-l volume, but none should be forced to do so.

3. At 15-minute intervals, subjects will again empty their bladder and collect all of the urine. Record the volume. If there is sufficient volume, determine the **specific gravity** or **density** of the urine. The term specific gravity refers to the weight of a given volume of fluid. The more concentrated the fluid, the higher the specific gravity. This will give some indication regarding the overall concentration of the urine. Note the color and viscosity of the urine as well.

4. To determine specific gravity, fill a urinometer cylinder two-thirds full with fresh urine. Place the hydrometer in the urine, and make sure it is not held to the wall of the urinometer by surface tension (Figure 12.5). Spinning the hydrometer when putting it into the urinometer may keep the hydrometer off of the wall; urinometers also may have ridges along the wall to help prevent this problem. Read the specific gravity at which the fluid level crosses the hydrometer's narrow tube. Be sure you understand the scale and markings on the hydrometer.

5. After the 45- or 60-minute collection, half the volunteers should exercise for 5 minutes. This exercise should be sufficient to elevate their heart and breathing rates.

6. Collect the urine on schedule at the next 15-minute time. Note any change in the exercise group compared with the control group.

7. Make a graph, with time on the x-axis and either Total Volume/15 Minutes or ml/min on the y-axis. What changes occurred in the total volume produced or in the rate of urine production in the exercise group?

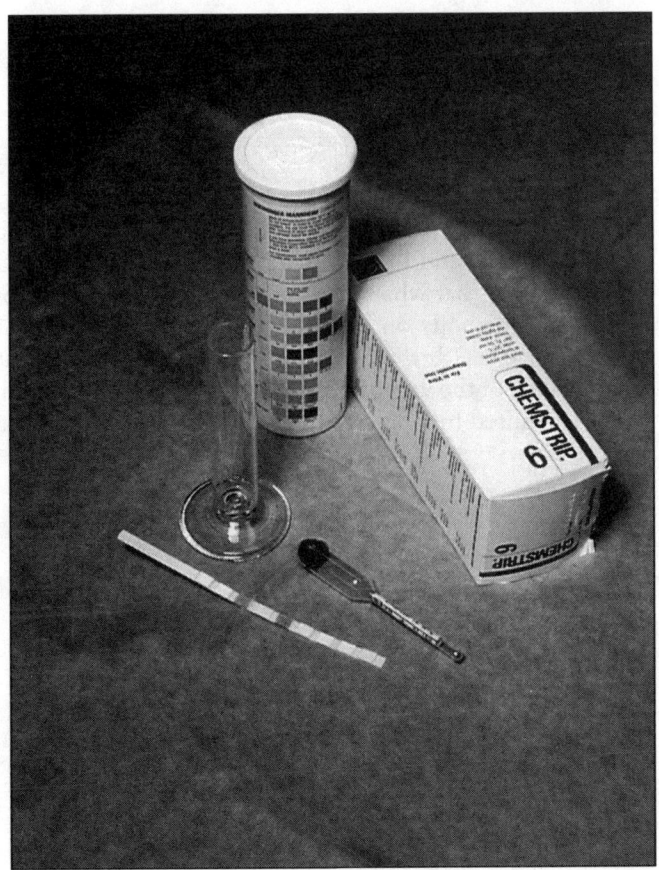

Figure 12.5
Hydrometer and Urinometer. To measure specific gravity, which relates directly to the total concentration of the urine, fill a urinometer with urine, then float a hydrometer in the cylinder. Spinning the hydrometer will keep the hydrometer from sticking to the wall because of the surface tension of the urine.

8. Add all of the individually timed sample volumes after the last sample has been collected. How much of the 1-l volume of distilled water consumed at the beginning has appeared as urine output?

QUESTION: Urine formation requires adequate blood flow and normal kidney function. What do your results tell us about the effects of exercise on urine formation?

Observations:

Conclusion:

QUESTION: If a hormone exists that is stimulated by exercise or loss of water from the body, how would that hormone affect formation of urine in the kidneys?

Observations:

Conclusion:

QUESTION: Some part of the 1 l of distilled water consumed at the beginning will have been collected by the end of the lab. Remember, however, that humans are in fluid-balance. This means that whatever fluid comes into the body on a daily basis will be eliminated from the body on a daily basis. How much of the 1-l volume was collected at the end of the lab? If there is quite a discrepancy, what does this indicate about the drinking habits of the volunteers? What other routes of fluid loss (besides urinary loss) must be accounted for to determine if a person is in fluid-balance?

Observations:

Conclusion:

Journal Note Normal kidney function is essential to the well-being of our body. Certain diseases, however, can cause kidney malfunction. What are some of these diseases? When the kidneys stop working, their role must be taken over by some artificial means. Kidney dialysis is used to keep such people alive, either while they wait for a transplant or for the rest of their natural lives. What are the different types of dialysis? Where is dialysis performed? Do you know anyone who is on dialysis? Compared with early methods, newer techniques now allow dialysis patients to live longer, and productively.

Analysis of Urine

During a routine medical examination, several types of samples may be taken from the body. Blood often is drawn to investigate the concentration of many materials, such as electrolytes, proteins, glucose, immunoglobulins, hormones, as well as red and white blood cells. Blood travels throughout the body, and it carries materials from one area to another. Urine also is a sample that can be collected to analyze what the body is eliminating, to better understand what it may still need, or to determine how well the kidneys are functioning.

There are protocols for artificial urine that allow students to analyze sterile, pathogen-free samples. Shmaefsky has provided information for this approach. (Shmaefsky, B. R. "Artificial urine for laboratory testing." *Am Biol Teacher* 1990; 52:170–172).

Analysis of Urine

Objectives:

- To appreciate the numerous characteristics of an easily collected bodily fluid.
- To understand the simple methodologies available for studying urine.

Materials:

- Latex gloves.
- Urine collection cups.
- Urinometer cylinders.
- Hydrometer.
- Wide-range pH paper (narrow-range pH paper, 6–8).
- Chemstrip (Boehringer Mannheim Diagnostics, Indianapolis, IN 46250), Urispec 9-way (Henry Schein, Inc., Port Washington, NY 11050), Multistix (Miles, Inc., Elkhart, IN 46515).

Procedure:

Caution: Urine is a bodily fluid; therefore, exercise care in its handling and disposal. Wear gloves when handling urine samples, avoid spilling urine, and wipe up any spills that do occur. Do not dispose of large quantities of urine in the sink. Clean glassware thoroughly, or place it in the special wash containers provided.

---

1. Collect a sample of urine, and analyze it based on the following criteria. Refer to Table 12.1 for normal characteristics.

   Color: Urine typically is yellow because of urochrome, which is a pigment derived from the breakdown of hemoglobin. Other colors can indicate problems within the urinary system. Certain foods may color the urine as well; for example, asparagus gives a green tint to the urine. Brown may indicate the presence of blood. Actual,

Table 12.1 Characteristics of Urine

Volume:

Adult, 1000–1500 ml/day
Obligatory output (volume necessary to rid the body of wastes), 350–400 ml/day
Polyuria = urine output > 2500 ml/day
Oliguria = urine output < 500 ml/day
Anuria = urine output < 100 ml/day

Specific Gravity:

1.016–1.022; range, 1.001–1.040

pH:

5.0–6.5, but may range from 4.5–8.0
Acidic pH on a normal, mixed diet
Basic pH on a vegetarian diet
More acidic pH in the morning

Color:

Pale to deep yellow (urochrome, a product of hemoglobin degradation gives color to the urine)
Urine usually is darker in the morning because of decreased water excretion during the night

Protein:

None
Exercise proteinuria is the occurrence of protein in the urine of athletes, especially runners and boxers

Glucose:

None
Glucosuria is most commonly caused by a blood glucose concentration that is too high (hyperglycemia)

Urea:

2 g per 100 ml
Derived from the breakdown of protein and, therefore, varies with the amount of protein in the diet

Ammonia:

Varies with acid excretion

fresh or frank blood in the urine should be visible as red material. The darker shade of any color indicates concentrated versus diluted urine. You probably have noticed already that vitamins and some drugs can alter the color of urine. Have you ever noticed that any other foods affect the color of your urine?

Transparency: Urine typically is clear. The presence of floating matter indicates infection or damage to the lining of the system. Cloudy urine indicates infection in some part of the system.

Odor: Freshly collected urine usually does not have an unpleasant smell. After sitting a while, however, it will develop an ammonia odor, because of bacteria breaking urea into ammonia. Some foodstuffs will affect the odor of urine as well. Diabetes mellitus often is diagnosed because of the fruity or acetone smell associated with the urine of patients with uncontrolled diabetes. Have you ever noticed that any other foods affect the odor of urine?

pH: Urine usually is slightly acidic. The pH can vary from very acidic to just slightly basic (4.5–8.0). Again, foodstuffs may affect the pH.

Specific Gravity: Another way to determine the urine concentration is to measure its specific gravity. This can range from 1.001 to 1.035.

Chemical Composition: There can be an array of different chemicals in the urine of any individual at different times and under different conditions. Urine contains urea, sodium, potassium, phosphates, sulfates, creatinine, and uric acid as the most common substances. Many other compounds may be present as well. Glucose, protein, blood cells, and bile pigments should not be present under normal conditions.

Note: There are numerous wet-chemistry methods to analyze many organic and inorganic constituents in urine. One method that has replaced many of these wet-chemical tests, at least for the qualitative identification of constituents, is the use of various dipsticks. These dipsticks are plastic strips with chemically treated sections attached that when moistened with urine and allowed to develop will change color. The color then can be compared with a chart on the side of the bottle and indicates the presence or absence of a particular material. There are individual dip sticks that indicate only one substance or combination strips that can be used to monitor many constituents at once.

---

2. Observe the color, transparency, and odor of the urine. If this is your first time observing urine samples quantitatively, it may be helpful to compare your sample with others in the class. This will help you to better understand some of the differences described earlier.

3. Obtain a dispenser containing wide-range pH paper. Remove a piece of pH paper, and dip it into the urine sample. Make sure the strip is thoroughly moistened. Then compare it with the color chart on the dispenser. Narrow-range pH paper also may be available; if so, try to determine more precisely the pH of the urine. (Combination dipsticks also may determine pH.)

4. Fill a urinometer cylinder two-thirds full with fresh urine. Place the hydrometer in the urine, and make sure it is not held to the wall of the urinometer by surface tension. Spinning the hydrometer when putting it into the urinometer may keep the hydrometer off of the wall; urinometers also may have ridges along the wall to prevent this problem. Read the specific gravity at which the fluid level crosses the hydrometer's narrow tube. Be sure you understand the scale and markings on the hydrometer.

5. Obtain a combination dipstick such as a Chemstrip, Urispec 9-way, or Multistix. Do not handle the chemically treated section of the strip. Follow the directions for dipping the strip into the urine. After the appropriate time, compare the strip as directed to the color chart on the bottle or box. Note there is an expiration date on the bottle: not only can the chemicals on the strip expire with age and exposure to heat and light, the color chart on the bottle or box can fade with such exposure as well.

6. Compare the results of the urine analysis with the list of characteristics and the results of the combination dipsticks. (Urine also may contain numerous floating or suspended materials, such as blood cells, squamous epithelium cells, and crystals. High-power microscopes and staining may be necessary to observe all the materials that can be in urine.)

QUESTION: Urine often is collected from athletes before competition, potential employees before employment, and in certain companies, from employees on an unannounced basis. What substances are being looked for in the urine? How do substances in the bloodstream actually get into the urine?

Observations:

Conclusion:

Journal Note Have you ever noticed the smell of your urine and how it might change throughout the day or after certain foods are consumed? Does your urine smell stronger in the morning or later in the day? What does physical exertion do to the appearance and odor of urine? Have you ever noticed that particular foods color or cause a distinct odor in your urine?

Section 13

The Reproductive System

Note to the Student

The reproductive system is the set of organs that allow the human species to propagate. Thus, the differences between "male" and "female" is the "stuff" of the reproductive system. Because the male and female systems are distinctly different, each is discussed separately. Each reproductive system contains the same basic architecture of gonads (i.e., testes in males, ovaries in females), connecting tubes (i.e., vas deferens and urethra in males, fallopian tubes, uterus, and vagina in females), and associated glands (i.e., prostate and seminal vesicles in males, mammary and mucous glands in females). The male system also has a specialized structure for sperm transfer, the penis, and the female system has a receiving structure for the penis, the vagina. A general note, the male system is the more complicated anatomically, but the female system is more complicated physiologically. This relates to the monthly cycle of the female system in preparation for pregnancy if fertilization occurs.

Meiosis: Reduction Division

This section reviews the anatomy of the male and female reproductive systems and the general aspects of their reproductive functions. Mitosis, which was mentioned in Section 2, is process the general body cells or somatic cells undergo as they divide into two identical daughter cells. This is how old and dead cells are replaced in the natural course of aging and growth.

The specialized sex cells known as germ cells or gametes (i.e., that is, the egg and the sperm), however, undergo a different process, which is called **meiosis,** before fertilization. **Fertilization** is the union of the sperm and the egg to form the zygote, which then develops into an embryo. The egg and sperm each must contain only half the number of chromosomes in a somatic cell, so that on fusion, the normal number of chromosomes in a cell is re-established. Meiosis reduces the chromosome number from 46 (i.e., that of the somatic cell) to 23 (i.e., that of the germ cell). Hence, the process is known as reduction division.

Furthermore, the process of meiosis differs bteween men and women. In men, meiosis results in four sperm; in women, meiosis results in only one egg. The goal, however, is the same: the number of chromosomes per sperm and egg is reduced by half. The process of sperm formation is known as **spermatogenesis,** whereas the process of egg, or oocyte, formation is known as **oogenesis.** Spermatogenesis occurs in the male gonads, or the testes, and oogenesis occurs in the female gonads, or the ovaries. The steps of each process are reviewed later.

Meiosis

Objectives:

- To understand the process of meiosis and its role in the production of sperm and eggs.
- To differentiate the processes of meiosis, or reduction division, and mitosis, or cellular division.

Materials:

- Microscope.
- *Ascaris* eggs.

Procedure:

Though we are interested in the process of meiosis in humans, it is impossible to see all of its stages, either in prepared testis or ovary slides. Therefore, one of the classic materials used to observe these stages is *Ascaris* eggs.

1. Select a slide, and place it on the microscope. Observe the stages of meiosis by scanning the slide and picking out the representative stages (Figure 13.1).

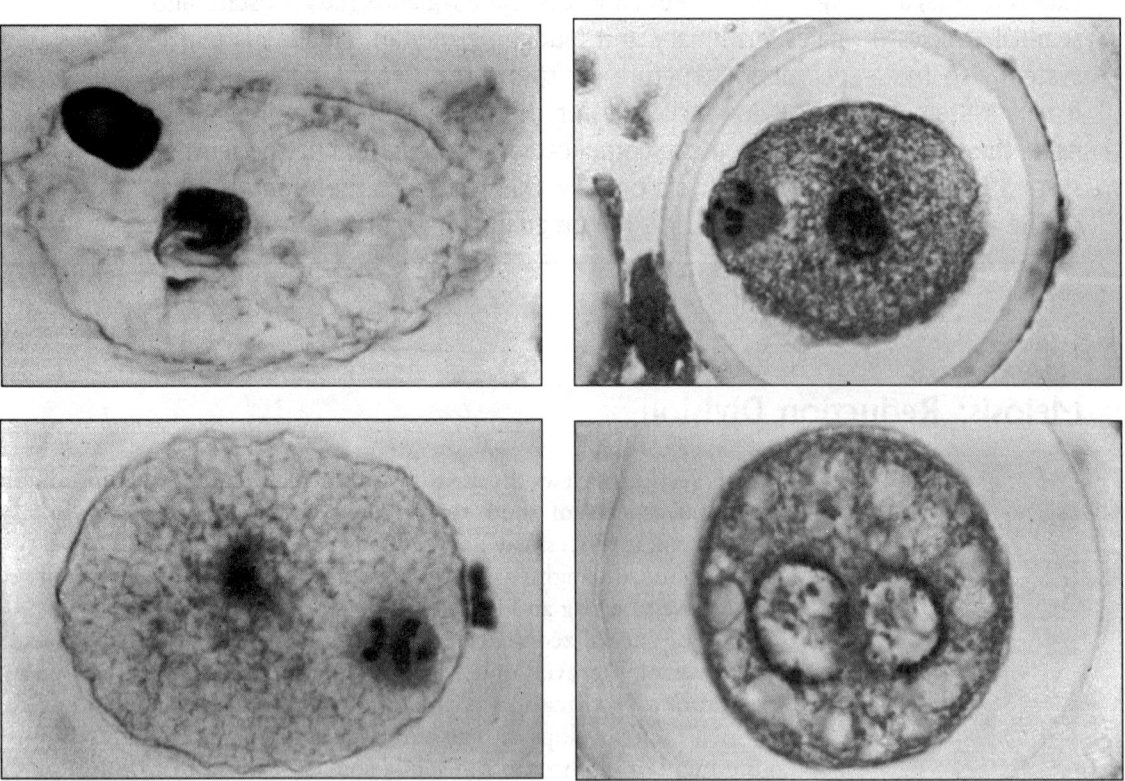

Figure 13.1
Ascaris Meiosis.
This series of photographss shows the stages of meiosis as it occurs in the worm *Ascaris*. Egg meiosis does not begin until the sperm has penetrated the egg; therefore, all subsequent stages will contain the male pronucleus.

2. Meiosis actually consists of two processes of division. These are referred to as meiosis I and meiosis II, and each contains the same stages previously associated with the process of mitosis.

Demonstrating of the process of meiosis as described above is a difficult process. An alternate method is to draw, or even construct, the representative stages of meiosis.

Meiosis Model

Objectives:

- To understand the process of meiosis and its role in the production of sperm and eggs.
- To differentiate the processes of meiosis, or reduction division, and mitosis, or cellular division.
- To understand meiosis by constructing its representative stages.

Materials:

- Text or reference.
- Paper plates.
- Markers.
- Pipe cleaners, three colors.

Procedure:

1. In your text or reference, find the section that describes meiosis. (Do not concentrate on differentiating the processes of sperm and egg production.)

2. With the paper plate as a cell and its edge as the cell membrane, use sections of pipe cleaners to represent chromosomes (Figure 13.2). To ensure you understand the process, use a minimum of three chromosomes, and use a marker to draw a nuclear membrane in the cell where appropriate.

Figure 13.2
A Paper-Plate Cell.
This photograph should give some indication of how to use the paper plates, pipe cleaners, and markers to make a cell model to illustrate meiosis.

Meiosis: Reduction Division

3. Construct or illustrate the following stages of the mitotic process:

   Prophase I

   Metaphase I

   Prophase II

   Metaphase II

   End of Telophase II

4. Explain the process to a neighboring group as you compare your layout with theirs. Discuss any differences, and determine if these differences are substantive.

QUESTION: This procedure calls for using the same-sized paper plate (e.g., dinner plates) for each stage in the process. Because the actual process includes little time for cell growth, would it make sense to use smaller plates (e.g., salad plates) for the later stages?

Observations:

Conclusion:

QUESTION: This layout displays meiosis in general, not spermatogenesis or oogenesis, but there are some important differences. How could you use different-sized (e.g., dinner, salad, and dessert plates) to help illustrate differences in the outcome of spermatogenesis and oogenesis?

Observations:

Conclusion:

The Male Reproductive System

Because the reproductive purpose of men is to produce sperm to fertilizing eggs produced by women, our discussion begins with the gonads, or **testes** (singular, *testis*) (*testis* (L), a witness), which are the organs that produce sperm (Figure 13.3). The testes are paired, oval organs approximately 5 cm in length that are contained in a sac-like extension of skin, the **scrotum** (*scrotum* (L), a pouch), in the pelvic region. The testes are located where the scrotum extends from the torso between the legs. Each

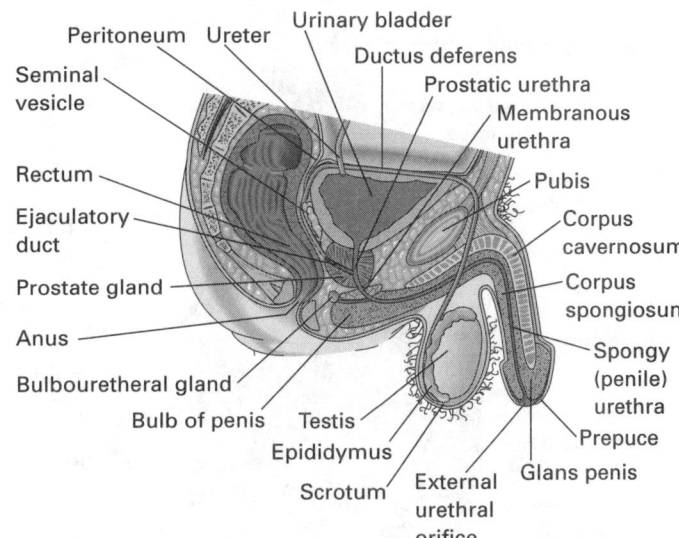

Figure 13.3
The Male Reproductive System.
This sagittal section of the pelvic region of the body shows the parts of the male reproductive system.

testis is covered and protected by a connective tissue sheath, the **tunica albuginea,** that also extends into the testis to provide partitions, which subdivide the organ into many lobules or compartments. Each lobule contains two or three **seminiferous tubules,** which are highly coiled and fill the lobule space. A microscopic cross-section of a seminiferous tubule reveals the various stages that occur in the process of **spermatogenesis,** or sperm formation.

Stages of Spermatogenesis

Sperm cells become capable on ejaculation of fertilizing eggs. Because sperm must leave the male body and face many hazards along the route toward insemination, it is necessary for huge quantities to be ejaculated to ensure fertilization.

Stages of Spermatogenesis

Objectives:

- To understand the stages of sperm formation.
- To appreciate both the germ-cell production and endocrine function of the testes.

Materials:

- Microscope.
- Testis slide, cross-section.

Procedure:

1. Select a slide showing a cross-section of the testis (Figure 13.4). Place it on the microscope, and examine it. Two cell types are found in the tubules: germ cells in various stages of development, which serve to produce the sperm, and supportive cells, which support and maintain the sperm cells.

Figure 13.4
The Testis.
(a) This cross-section of the testis shows the many seminiferous tubules that make up the organ. (b) This detailed cross-section of a single seminiferous tubule shows the stages of spermatogenesis.

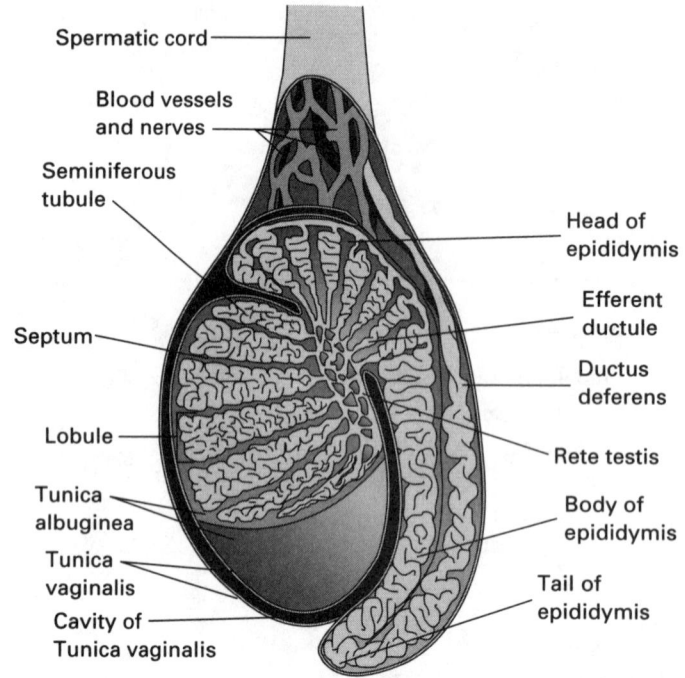

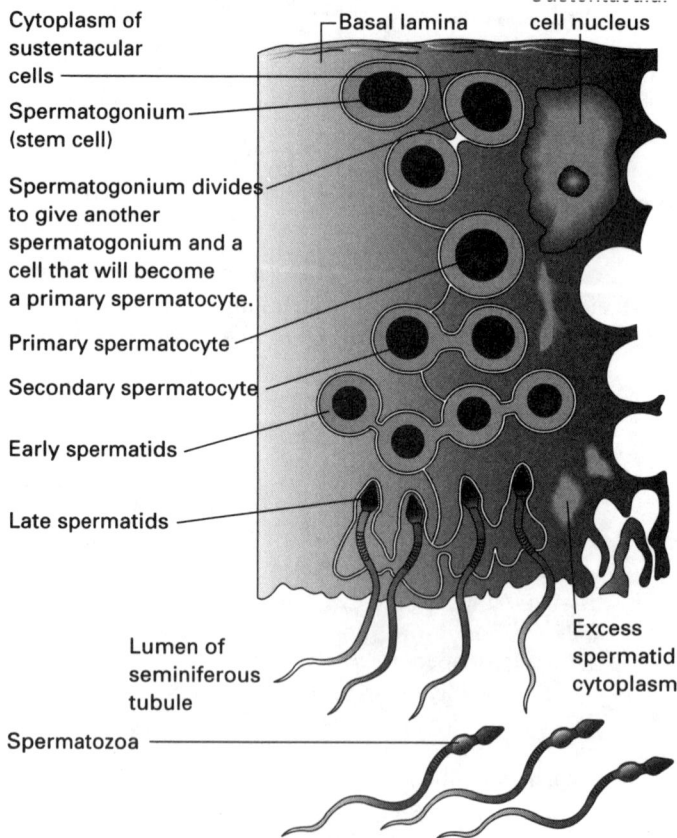

218   SECTION 13   The Reproductive System

2. Identify the following cells within a seminiferous tubule:

   Spermatogonia (singular, *spermatogonium*): This precursor cell will give rise to the cells that become sperm. These cells divide by mitosis, but with each division, a new daughter spermatogonium is formed as well as a cell that will go on to develop into sperm cells. The spermatogonia are located along the outer wall of the seminiferous tubules.

   Primary spermatocytes: These cells will divide meiotically to produce the remaing stages of sperm development. These cells still have the full complement of chromosomes for a human cell (i.e., 46). This number of chromosomes is referred to as the "2n" number. When they divide, these cells give rise to two new, secondary spermatocytes.

   Secondary spermatocytes: These cells are the product of the first meiotic division. As meiosis is reduction division, each secondary spermatocyte has half the normal number of chromosomes for a human cell (i.e., 23). This number is referred to as the "n" number of chromosomes. Each secondary spermatocyte enters the second meiotic division to give rise to two new spermatid cells.

   Spermatids: These cells are ready to become functional sperm cells, but they are not fully mature or viable at this point. In other words, they are incapable of fertilizing an egg. Just as spermatogonia line the wall of the tubule, the other cells are arranged in layers along the lumen of the tubule. In the lumen itself are many developing and maturing sperm, or spermatozoa.

   Spermatozoa (singular, *spermatozoan*): These cells are the fully developed sperm. They still are not viable, but the maturation process will occur as they make their way through the ducts of the male reproductive system. As these various stages develop in the tubule, the sperm move further away from the body's blood supply and are nurtured by the Sertoli cells.

   Sertoli (nurse) cells: These cells surround and support the various developing stages of sperm production. While it is difficult to see on the microscope slide, they are actually rather large cells that literally surround the developing cells.

3. After examining the contents of the seminiferous tubules, note that in the cross-section, the tubules are circular and have spaces between them. In these spaces are the **interstitial** or **Leydig cells,** which produce testosterone as the principal sex hormone of the male body.

Coming from each lobule are the **straight ducts,** which then connect to a meshwork of ducts called the **rete testis** (*rete* (L), a net, network). The next duct is the **efferent ductule,** which empties into the **ductus epididymis,** which actually is contained in a structure attached to the posterior aspect of the testis called the **epididymis.** The epididymis is described as a comma-shaped structure that is made up of a head, a body, and a tail. In the epididymis, sperm become motile, or capable of self-propulsion, and, therefore, of fertilization. Sperm enter the head of the epididymis and flow from the tail into the **ductus deferens** or **vas deferens** (plural, *vasa deferentia*), which carries the sperm into the pelvic cavity, posterior to the bladder, where the duct turns inferiorly to connect to the urethra. Just before the vas connects to the urethra, it enlarges to become the **ejaculatory duct.** As mentioned in the previous section, the urethra carries both the sperm for reproduction but also the urine for the urinary system. The opening of the urethra is called the **external urethral orifice,** and the sperm exit the male reproductive system through this opening.

The testes are supported in the scrotum by the sac-like structure of the scrotum itself and also by a "life line" known as the **spermatic cord** (Figure 13.5). The spermatic cord is a bundle of tubes that connect the testes with the body cavity, and it consists of the testicular artery and veins, nerves, lymph vessels, cremaster muscle, con-

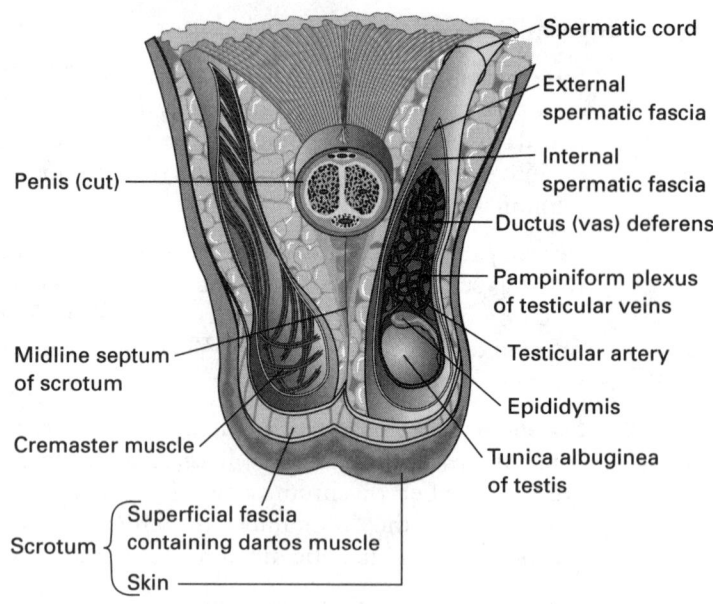

Figure 13.5
The Spermatic Cord.
Each testis is attached to the body by the spermatic cord. The cord is a bundle of arteries, veins, nerves, lymph vessels, and muscles that support and nourish the testis.

nective tissue, and ductus deferens. The functions of these various components are obvious, except for that of the **cremaster muscle** (*cremaster* (G), a suspender). The cremaster muscle helps to maintain proper temperature for sperm formation, either by relaxing to let the testes fall further away from the body cavity and, thus, provide a lower temperature or by contracting to bring the testes closer to the body during colder seasons of the year. Sperm formation requires temperatures several degrees cooler than body core temperature. To assist in maintaining this lower temperature, veins returning from the scrotum are arranged in the **pampiniform plexus,** which is a heat exchange structure that removes heat from the warmer arterial blood so that the testes remain cooler.

Glands of the Male Reproductive System

Three glands support the male reproductive system. The pair of **seminal vesicles,** each of which is shaped like a cluster of smaller, sac-like glands, are located posterior to the urinary bladder, and they are connected to and drain into the ductus deferens where this organ meets the ejaculatory duct. Each seminal vesicle is associated with either side of the body. A single, walnut-sized **prostate gland** surrounds both the urethra where it exits the urinary bladder and the ejaculatory ducts at the point where they connect to the urethra. Finally, a pair of pea-sized **bulbourethral glands,** or **Cowper's glands,** drain their secretions into the urethra just below the prostate gland. Collectively, these three glands produce secretions that provide nutrition for the sperm, chemicals to increase their motility, and lubrication for sexual intercourse.

QUESTION: These three glands actually provide the bulk of the semen that is ejaculated during intercourse. Use your text to differentiate the volume and contents that each gland contributes and what purpose is served by each secretion. (The largest volume of secretions comes from the seminal vesicles. It contains the sugar fructose. What purpose would this sugar play?)

Observations:

Conclusion:

External Male Genitalia

Besides the scrotum, the **penis** also is part of the external male genitalia. A tool for placing the sperm-containing semen into the vagina of the female, the penis essentially is a shaft-like appendage to the lower anterior abdominal wall, which is covered with skin and contains the urethra and three masses of erectile tissue (Figure 13.6). The bulk of the penis is a pair of erectile tissue masses, the **corpora cavernosa** (singular, *corpus cavernosum*) (*corp* (L), a body; *cavern* (L), a cave, chamber), that make up the dorsal and lateral aspects. The other mass is the **corpus spongiosum,** which is on the ventral side and contains the urethra. All three masses of erectile tissue extend into the base of the abdominopelvic cavity floor, where they anchor the penis into the body.

QUESTION: If the erectile masses in the penis were only in the penis and did not extend into the body cavity, could the penis still become hard and rigid? Could it become erect (i.e., stand perpendicular to the body)

Observations:

Conclusion:

Figure 13.6
The Penis.
This cross-section shows the internal structure of the penis. Note the bundles of erectile tissue, the corpora cavernosa, and the corpus spongiosum.

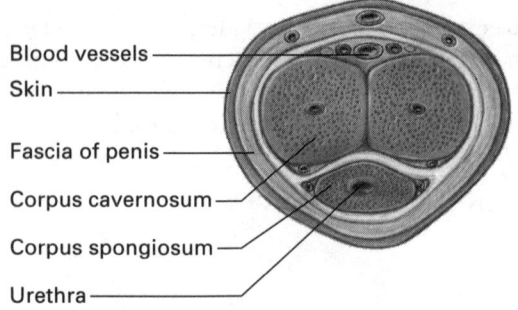

The corpus spongiosum expands at the free end of the penis to produce a cap-like structure called the **glans** (*glans* (L), an acorn). The firmness of the glans allows the many nerve endings in it to be stimulated during sexual activity. The glans in a male newborn is surrounded by a protective skin, the **prepuce** or **foreskin.** The prepuce often is removed surgically soon after birth because of religious, cultural, or medical tradition; this procedure is known as **circumcision.** In an uncircumcised man, glandular secretions accumulate in the space between the glans and the prepuce as a substance called **smegma** (*smegma* (G), a soapy secretion).

QUESTION: Third-degree burns over extensive areas of the body are serious medical situations. Artificial as well as cadaveric skin offer benefits to such patients by providing a temporary, protective covering over burned areas. Natural skin grown in tissue culture also improves the survival of burn patients, and foreskin removed from newborns is being used to provide skin for tissue culture. Why would this tissue be good for culturing?

Observations:

Conclusion:

Erection and Ejaculation

Before sexual arousal, the penis is flaccid. When there is visual, emotional, or mental stimulation of a sexual nature, however, certain autonomic nervous system reflexes cause an erection. Parasympathetic fibers to the arterioles supplying the penis result in vasodilation, which is the widening of blood vessels. Thus, these vessels allow more blood to flow into the erectile tissues than normal. As the erectile tissues fill with blood, they compress the veins, which results in more blood flowing into the penis than can leave it. This process causes **erection.** If stimulation continues, sympathetic outflow from the autonomic nervous system eventually causes **ejaculation.** In this process, smooth muscle in the ducts and skeletal muscles at the base of the penis contract, and the semen is ejected from the urethra. **Orgasm** is the climax of the sexual act. The material emitted is called the **ejaculate** and usually has a volume of 3.5 ml. On average, there are 80 million sperm per 1 ml of ejaculate and, therefore, 280 million sperm per ejaculation.

QUESTION: The events associated with orgasm are more complicated than simply those listed earlier. Signals are sent to various parts of the body to guarantee that all processes occur in the proper sequence. Sperm have two paths available as they pass from the ejaculatory duct to the urethra. What muscle contracts to ensure that sperm flow out of the penis and not into the bladder?

Observations:

Conclusion:

QUESTION: We often hear that it only takes one sperm to fertilize an egg. That there are 280 million or more sperm in the typical ejaculate, therefore, seems a bit like overkill. Can you hypothesize why so many sperm are present when only one will fertilize the egg to produce a zygote? (The condition of multiple sperm fertilizing an egg is called polyspermy, which results in miscarriage.)

Observations:

Conclusion:

Anatomy of a Sperm Cell

A sperm cell is distinctive in several ways. It is the only cell, in natural situations, that is transferred from one individual to another as part of sexual intercourse. It also is the only flagellated cell in human anatomy; therefore, there are no flagellated cells in human females. Both genders do have ciliated cells, especially in the respiratory system to move mucus along the airways, and both flagella and cilia have the same basic internal structure. They differ, however, in their length and numbers. Cilia are short and numerous, whereas flagella are long and singular. Both move the cells to which they are attached or a material across the surface of those cells.

The sperm cell essentially is a nucleus and a tail (Figure 13.7). Its role in sexual activity is to convey the male genetic contribution to the egg, which contains the female genetic contribution. The sperm cell has a **head,** a **neck,** a **midpiece,** and a **tail** or **flagellum.** The head contains the nucleus and an **acrosomal cap,** which is a structure that contains enzymes to assist the sperm in penetrating the egg. The neck simply is the connection between the head and the midpiece, which contains the many

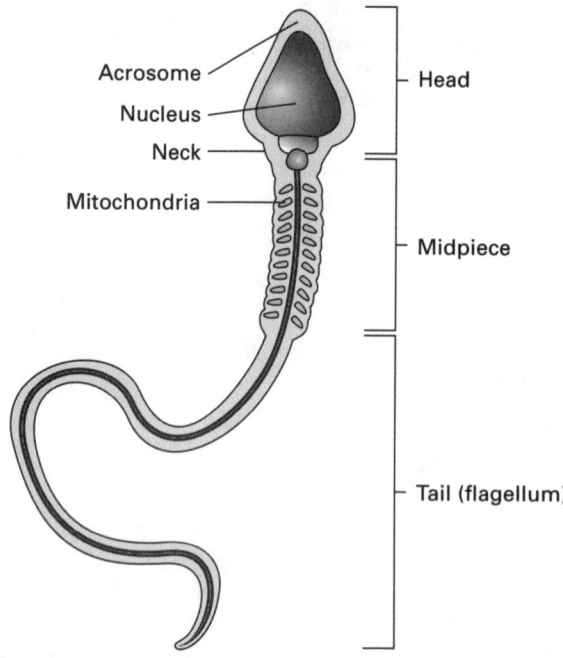

Figure 13.7
A Sperm Cell.
There are three regions in human sperm: the head, the midpiece, and the tail.

mitochondria that provide the power (in the form of adenosine triphosphate [ATP]) to propel the sperm toward its target. Finally, there is the flagellum, which is a whip-like structure that is capable of propelling the sperm cell from 1 to 4 mm/min.

The Female Reproductive System

During embryonic development, **mesodermal** (*meso* (G), the middle; *derma* (G), skin) tissue along the ventral abdominal cavity wall develops into the gonads of the appropriate gender. Testes descend into the scrotum in men; ovaries descend through the lower abdominal cavity and into the pelvic cavity in women. The ovaries remain inside

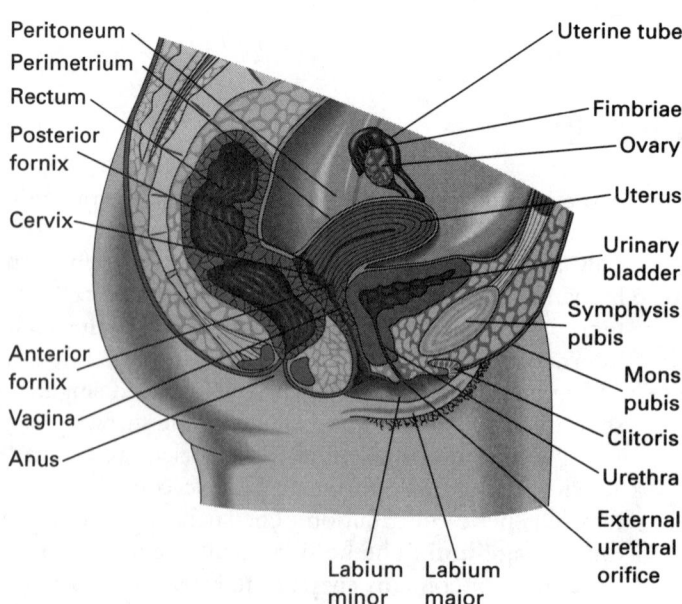

Figure 13.8
The Female Reproductive System.
This sagittal section of the pelvic cavity shows the organs of the female reproductive system.

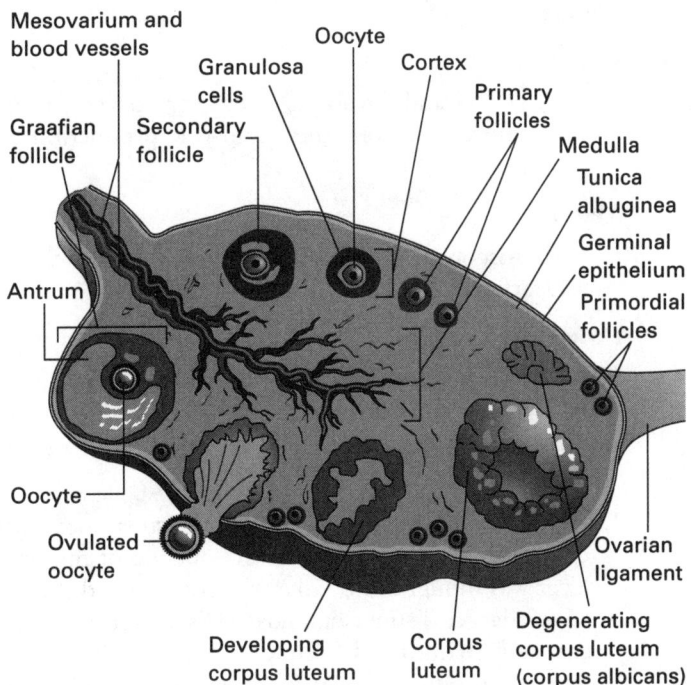

Figure 13.9
The Ovary.
The central portion of the ovary containing connective tissue and blood vessels is the stroma; the outer portion containing the different stages of follicles is the cortex. The different stages of oogenesis can be observed in the ovary.

the body associated with the other organs of the female reproductive system (Figure 13.8.) The ovaries are held in place by various ligaments, and connect with the uterus by tubes for the purpose of releasing eggs. The uterus opens to the outside of the body by way of the birth canal. The ovaries themselves are paired, oval organs 3.5 cm in length and 2 cm in width (Figure 13.9). The ovaries can be divided into an **outer cortex,** where many of the stages of egg development can be found, and an **inner medulla,** where the blood vessels, lymph vessels, and nerves supplying the ovary are contained. Microscopic examination of ovary cross-sections reveals the various stages of egg formation, or **oogenesis.**

Stages of Oogenesis

Just as the male's gonad, the testis, produces sperm, the female's gonad, the ovary, produces eggs. Both processes require the primitive germ cells to undergo meiotic division to attain the correct number of chromosomes.

QUESTION: A man produces sperm by the millions, but a woman produces very few eggs. As mentioned, even in meiosis, four sperm are produced, compared with only one egg. Why is this so? What are the differences in the size and functioning of sperm and eggs that relate to this question?

Observations:

Conclusion:

Stages of Oogenesis

Objectives:

- To understand the stages of egg, or oocyte, formation.
- To appreciate both the germ-cell production and endocrine function of the ovaries.

Materials:

- Microscope.
- Ovary slide, cross-section (mammalian).

Procedure:

1. Select a cross-section slide of the ovaries. Place it on the microscope, and examine it. Two cell types are found in the cortex: germ cells, and follicle cells. Together, these cell types make up the operative units of the ovaries, the **follicles** (Figure 13.9).

2. In addition to the thin peritoneum of squamous epithelial cells, there also is a layer of **germinal epithelium** on the outside of the ovaries. When it was named, this layer was thought to be that which contained the ova. This was because of the relatively large, distinct cuboidal cells within it. Immediately inside this epithelium are many **primordial follicles,** which are the reserve stock of germinal cells of the female reproductive system. They represent all the potential ova that a woman will ever produce in her life. A primordial follicle contains a **primary oocyte;** therefore, a follicle is a structure made of follicular cells and an immature ovum.

QUESTION: This section contains an important distinction between male and female reproductive physiology. At birth, females already contain all the eggs they will ever produce (approximately 700,000). Compare this with the situation in males.

Observations:

Conclusion:

3. There also will be some follicles with enlarged cells. These are the **primary follicles,** which are follicles in the process of dividing and developing. One of these developing follicles will be the one that ovulates.

4. The follicle cells divide, and the oocyte grows. A fluid-filled space, or **antrum,** develops. At this point, the same follicle is called a **secondary follicle.**

5. The growth of cells and enlargement of the antrum continues, thus causing the follicle to bulge from the ovary surface. This process is referred to as maturation, and such a follicle is called a **mature follicle.** When the follicle is just under the ovary surface and significantly increased in size, it is called a **Graafian follicle.**

6. The Graafian follicle ruptures, and the oocyte, or ovum, is expelled or ovulated into the body cavity near the ovary, where it will be swept into the **fallopian tube,** or **oviduct,** by the cilia. The follicle that released the oocyte then transforms into a pale-colored, glandular structure called the **corpus luteum** (*lute* (L), yellowish).

Figure 13.10
The Fallopian Tubes, Uterus, and Vagina.
These accessory organs—the fallopian tubes, uterus, and vagina—function with the ovaries to complete the female reproductive process.

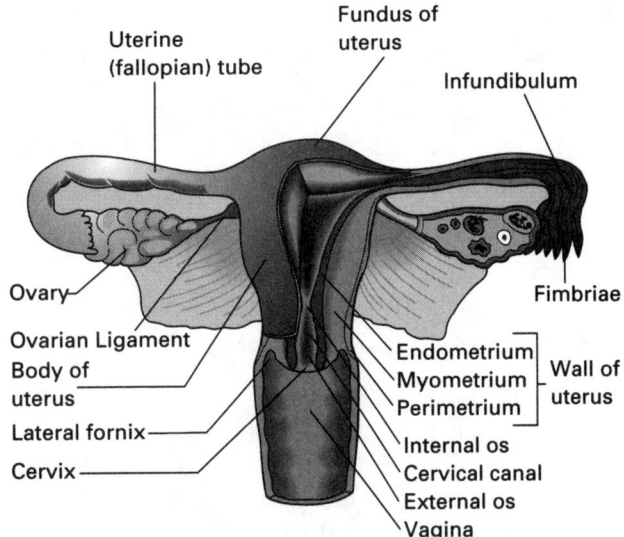

The corpus luteum produces hormones for uterine development and maintenance during the early months of pregnancy. If no pregnancy occurs, the corpus luteum will disintegrate.

The opening of the fallopian tubes, through which the ovum enters to proceed to the uterus, is called the **infundibulum** (*infundibulum* (L), a funnel) (Figure 13.10). There are **fimbriae** (*fimbria* (L), a fringe), or finger-like projections, surrounding the infundibular opening and that are covered with cilia to assist the ovum into the **oviducts** (*ovi* (L), an egg; *duct* (L), lead). The oviducts connect with the **uterus** (*uterus* (L), the womb), which is a midline, sac-like structure with a wall containing smooth muscle (i.e., **myometrium**) and that is lined by a mucous membrane that builds up monthly (i.e., **endometrium**). The lining of the uterus is sloughed off each month if fertilization does not occur during the **menstrual cycle.** The menstrual cycle refers to the approximate monthly build up of uterine lining in preparation for pregnancy, followed by loss of the lining if the ovum is not fertilized. The main part of the uterus is referred to as the **body,** and it narrows inferiorly to the **cervix** (*cervix* (L), the neck). The opening of the cervix is the **cervical canal.**

The cervical region of the uterus projects into the **birth canal,** or **vagina** (*vagina* (L), a sheath). Where the cervix projects into the vagina, there is a depression around it, which is called the **fornix.** The vagina, which is the canal the penis is inserted into during sexual activity, is lined with stratified squamous epithelium for protection. The vagina also stretches considerably during childbirth. The opening of the vagina to the outside is called the **vaginal orifice.**

Birth Control Devices

Now that we have discussed and examined the the male and female reproductive systems, we examine methods of birth control. The following exercise involves consideration of different forms of birth control devices and how they work.

Birth Control Devices
Objectives:
- To become familiar with different forms of birth control devices.
- To understand how each device prevents pregnancy.

Materials:
- Birth control devices and literature, including:
  Condoms
  Birth control pills
  Diaphragms
  Sponges
  Intrauterine devices
  Spermicidal preparations
  Tubal ligation (drawing)
  Vasectomy (drawing)
- Models of the male and female reproductive tracts.

Procedure:

1. Examine all of the birth control devices that are available, and read and discuss the literature provided (Figure 13.11). Your text has an excellent discussion concerning a wide variety of birth control devices; it also looks at some new methods under development and in testing.

2. Group the devices into several broad categories based on the method by which they interfere with the reproductive process. Use the following to assist you in this grouping:

	Blocks Sperm	Blocks Eggs	Kills Sperm	Blocks Ovulation
Condoms	___	___	___	___
Birth control pills	___	___	___	___
Diaphragms/sponges	___	___	___	___
Intrauterine devices	___	___	___	___
Spermicidal preparations	___	___	___	___
Tubal ligation	___	___	___	___
Vasectomy	___	___	___	___

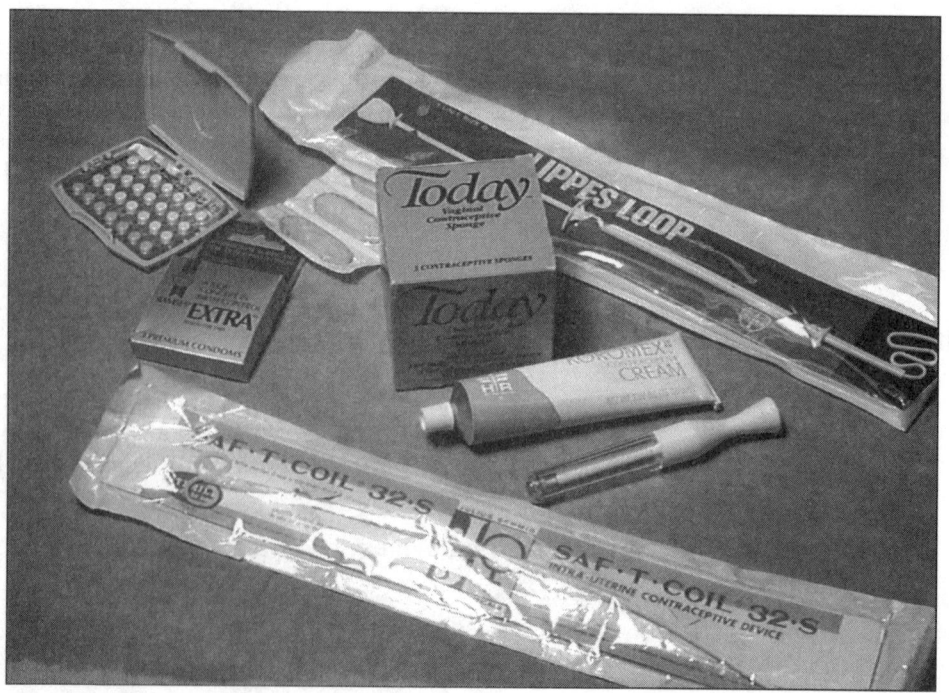

Figure 13.11
Birth Control Devices.
A variety of birth control devices currently are available in the United States.

Figure 13.12
The Vulva.
The collective term for the external genitalia of the female is the vulva.

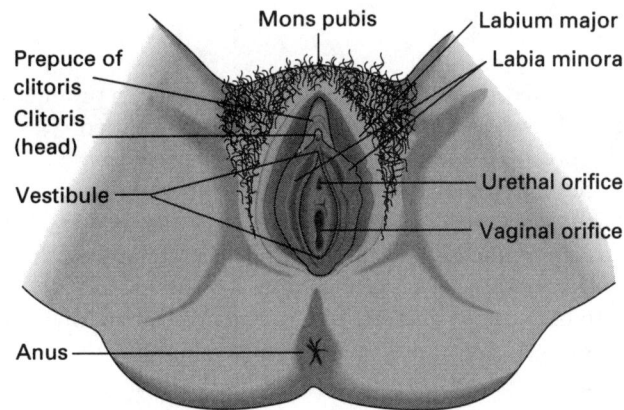

QUESTION: Barrier methods of birth control include the diaphragm, which prevents the entry of sperm into the uterus. A diaphragm is a cup- or dome-shaped, latex-rubber device that cups or covers the opening of the uterus. Where is an anatomically protected area to place the diaphragm?

Observations:

Conclusion:

The external genitalia of the female, which collectively are referred to as the **vulva** (*vulva* (L), a covering, wrapper) surround the vaginal orifice (Figure 13.12). The vaginal orifice is located in an anterior-posterior plane, between the urethral orifice and the anal opening. The area anterior to the anal opening and surrounding the vaginal and urethral orifices is called the **vestibule,** and it is bordered laterally by the **labia minora** (*labia* (L), a lip). The labia minora are fleshy "lips" or folds of mucous membrane that are richly supplied by glands. The anterior-most structure of the vulva is a fleshy elevation called the **mons pubis,** which is covered by pubic hair. Extending posteriorly from the mons are two larger "lips," or the **labia majora,** which are composed of skin and, therefore, are covered with pubic hair. At the point where the labia minora come together anteriorly is the **clitoris.** It is richly endowed with sensory nerve endings, and it provides stimulation to the woman during sexual activity. It also contains erectile tissue and is **homologous** to, or has the same embryonic origin as, the glans of the male body, just as the labia majora are homologous to the scrotum.

Mammary Glands   229

Mammary Glands

Other accessory structures of the female reproductive system are the **mammary glands** (*mamma* (L), a teat), which are modified sweat glands located in the **breasts.** (Figure 13.13). Both men and women have breasts, but only in women do they produce milk, which nourishes the offspring. Both genders have **nipples** on the breasts in the center of a pigmented area, the **areola** (*areola* (L), a little, open space). There are hair follicles associated with the areolas, and the roots of these hairs projecting from the skin cause the surface to be bumpy. The actual mammary glands are divided into between 15 and 20 lobes, which in turn are divided into lobules that contain clusters of glandular cells called **alveoli** (singular, *alveolus*). The alveoli secrete milk into a duct network that conveys the milk to the nipple.

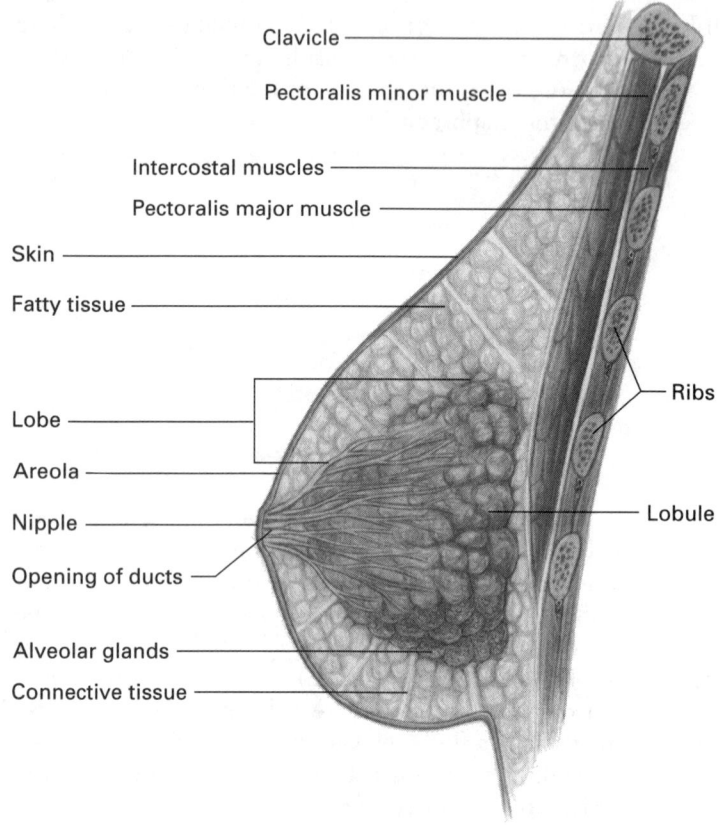

Figure 13.13
The Mammary Gland.
This sagittal section shows the large amount of fatty tissue in the female breast.

QUESTION: A woman's breasts become larger during pregnancy. A condition among men in which obvious breasts appear is called gynecomastia. That female breasts enlarge with pregnancy tells you what about the probable cause of male breast enlargement?

Observations:

Conclusion:

Pregnancy

The principal function of the reproductive systems in humans is fertilization of an egg, so it can develop into a new individual. All species have a drive to maintain themselves, and the human species is no different.

When an egg is surrounded by sperm, the sperm try to penetrate the outer, protective layers of the egg. Remember that millions of sperm enter the female reproductive tract during sexual activity. The acrosome of the sperm contains enzymes that help to penetrate the coverings of the egg, but when one sperm penetrates the egg, a fertilization membrane forms to prevent polyspermy.

An egg is viable for approximately 12 to 24 hours after ovulation. Sperm are fertile for 48 hours in the female reproductive system. Therefore, if intercourse occurs 48 hours before or 12 hours after ovulation, fertilization may occur. Because the fertilized egg takes 6 days to make its way down the fallopian tube to the uterus, fertilization must occur in the uppermost tubes and not anywhere near the uterus. The zygote (i.e., the fertilized egg) is carried down the fallopian tube by a combination of ciliary action and muscular contractions, until it finally arrives in the uterus. The process of **implantation** then begins, and it takes about one week to complete. The zygote has already started dividing, however, to develop into a new individual.

Main Features of Embryonic and Fetal Development

The first 2 months of development are referred to as embryonic. The term **embryo** is used to describe a developing human during the stages of early cell division, germ-layer development (i.e., formation of the ectoderm, endoderm, and mesoderm), and organogenesis (i.e., formation of the major organ systems). The term **fetus** is used to describe any organism after the embryonic stage, when the organism looks more like its own species rather than another. In humans, the fetal stage begins at approximately 8 weeks, and development past this time is called fetal. Table 13.1 provides a chronological sequence of embryonic and fetal development.

Table 13.1 Chronological Sequence of Major Changes in Development

Days 1–4

Early division occurs during movement of the embryo down the fallopian tube. If the first two cells separate, identical twins develop; two separate eggs are required for fraternal twins to develop. The embryo and extraembryonic cells number approximately 100 at this time.

Days 5–6

Outer feeding and supporting membranes (i.e., chorion, amnion, yolk sac, allantois) are forming rapidly and require most of the extraembryonic cells. The outermost cells are called the trophoblasts, and they assist in implantation to embed the embryo into the uterine wall.

Day 7

Implantation.

Days 8–14

Implantation continues.

1 month

The embryo is 5-mm long, and the heart is pumping. The limb buds are obvious and a tail evident. The digestive tract also is forming.

6 weeks

The embryo is 14-mm long. Cartilaginous skeleton is forming, and the head is a large, obvious structure. The limbs and the digits also are obvious, and a face and neck are forming. The tail is almost gone.

2 months

The embryo is 30-mm long. There is a flat nose and fused eyelids, and parts of limbs are evident. Digits are well formed, and bone formation has begun.

3 months

The fetus is 75-mm long, and the head is very obvious and large. Pinna are present, nails are forming, and tooth buds are beginning. Eyes are internally formed, but the lids still fused. Gender is apparent, and the heartbeat is readily detectable external to the woman's body.

4 months

The fetus is 17-cm long. There is an obvious human face and other features. Heartbeat is strong, and the fetus is moving. The skeleton is visible, and growth of the body is catching up with the head.

5 months

The fetus is 28-cm long, and the dome of the uterus is at the woman's navel. The eyelids still are fused, and some hair is on the head. Internal organs are nearly developed, but the lungs are not ready.

6 months

The fetus is 34-cm long. Nails and eyelashes are complete, and the eyelids are open.

7 months

The fetus is 39-cm long. Premature babies delivered at this age have a good chance of survival in sophisticated nurseries.

8 months

The fetus is 44-cm long. More growth is occurring, and the chance of survival is much greater.

9 months

The fetus is 51-cm long. Average birth weight is 7.0–7.5 pounds, with boys tending to be heavier.

Section 14

Genetics

Note to the Student

Often, the topics discussed in this section can be covered in different chapters of a textbook. There may be some coverage of cell division in chapters on the cell, and genetics can be discussed in the chapters on reproduction and development. There could be special chapters on genetics and genetic diseases. Few areas of human anatomy and physiology are as dynamic and rapidly changing as the field of genetics. Early studies dealt with such phenomena as pea color and texture inheritance, but today, the very molecules that control every feature of the human being are under study. The semester probably is nearly over as you prepare to enter this section. Many other aspects of your life may just be starting, and this material will be directly related.

Basic Principles of Genetics:

Your text provides considerable detail on this topic; however, we quickly review them here as well so that everyone be at a similar starting point. In each cell is an organelle called the nucleus. It contains the information that codes for all the characteristics of adult humans. For a human to develop, this information must be reliably copied and passed on to the two new daughter cells whenever cell division occurs.

Just think how many times this process of copying and transferring genetic information to the two daughter cells must happen. When an egg or an ovum is fertilized, the developing organism is a single cell. Initially, there is a doubling in cell number each time it divides. The single cell becomes two, and the two become four, the four become eight, and so forth. This continues through the morula, blastula, and gastrula stages and into the embryonic stages. As the cells differentiate, however, and begin having different roles and locations in the developing embryo, the rate of division for these cells changes. Some continue to divide rapidly, and others slow down.

QUESTION: The single cell divides until is becomes the very complicated multicellular organism we call an adult human. How many mitotic divisions are needed to reach the 60 to 100 trillion cells of an adult human body? You probably will be surprised by the answer. Remember, however, that all the cells created in these divisions do not survive, and that the body is constantly losing cells in the form of blood cells that wear out, skin and digestive-lining cells that are worn away, and the many other cells that are lost from the body. These cells must be constantly replaced. Still, it will be interesting to calculate the number of divisions, pretending that no cells are lost from the body, needed to reach the number mentioned.

Comments:

Conclusion:

The material in the nucleus that actually contains this information is deoxyribonucleic acid, or DNA. The **chromosomes,** which are composed of DNA and protein, generally are dispersed in the nucleus as **chromatin,** so that the contained information can be accessed for use by the cell. When the cell is ready to divide, however, the chromatin condenses into chromosomes for careful partitioning and transfer to the two daughter cells.

Information is carried on pairs of chromosomes, one of which comes from the father and one of which comes from the mother. Therefore, the genetic make-up of an individual (i.e., his or her **genome**) is derived equally from the father and the mother. There are two types of chromosomes regarding what they control. Most chromosomes are referred to as **autosomes,** which contain genetic information for mostly non–sex related functions. The other type are **sex chromosomes,** which control the proteins that relate to sexual development and function. Humans have 23 pairs of chromosomes, 22 pairs of autosomes, and 1 pair of sex chromosomes, for a total of 46 chromosomes. This number is referred to as the **euploid** number, or the **2n** condition.

QUESTION: The Human Genome Project is a worldwide effort to determine the sequence of information in every chromosome of an average human. What benefits do you think will accrue from this effort? How much do you think this information is worth? Do you see any direct benefits to you?

Comments:

Conclusion:

The genome often is referred to as the **genotype**, which is the actual genetic information contained within each cell's genetic material. Much of this genetic information is expressed early in development, as the various parts of the body form in the embryonic and the fetal phases. Some of this information may not be expressed until later in life, or it may not be expressed at all. In fact, quite a bit of genetic information never gets expressed at any time in the development of most species, including humans. The information that does get expressed, however, conveys particular characteristics on the individual. That peculiar set of characteristics as expressed from an individual's genome or genotype is called the **phenotype,** which is the outward expression of the genes encoded on the chromosomes. This outward expression is modified by the many variable environmental factors an individual is exposed to during development, both in the uterus and after being born. Hormones, nutrition, and even temperature can influence the direction of development; thus, an individual is formed to grow and meet the challenges of our environment.

How Do Genes Work?

Even the minutest detail of DNA copying and expression is being studied in countless labs, but here, we discuss the basic features of how information from the genome makes itself known in the phenotype. **Genes** are the units of information in the chromosomes that code for all the characteristics that make us human. The information coding for the characteristics may take different forms, and these are referred to as **alleles**. Our discussion focuses on situations in which there are only two alleles for any given gene in coding for a characteristic. Realize, however, that many characteristics are so complicated there are multiple alleles for each gene or multiple genes that code for a characteristic.

Alleles contain the information for cells to make proteins. The structure of these proteins is coded by the information in the alleles. Because there are two types of alleles for each gene, one most often has a stronger, or **dominant,** expression than the other, which is **recessive**. Some diseases or conditions are caused by recessive alleles coding for proteins that are not normal or are defective in function. To understand the basic mechanisms of how this information determines the characteristics, or **traits,** of an individual, people have observed how these traits are passed from parents to offspring. These observations have been going on forever, and to this day, even people with no understanding of genetics or genetic inheritance notice whether children look like their parents. Table 14.1 lists several disorders as well as normal traits that have a strong genetic basis. Many human genetics texts will have much more extensive listings.

Geneticists have agreed by convention to refer to the dominant gene of the trait being studied by an upper-case letter. The same lower-case letter then represents the recessive allele. Thus, when studying the inheritance of pigment colors, if the dominant color or condition is red and the recessive is yellow, the dominant allele would be denoted by R and the recessive by r. This technique is meant only to allow us to study the principles of inheritance; therefore it really does not matter which letters are chosen. Remember that each individual gets one of each pair of chromosomes from each parent, so that one of the alleles comes from the mother and the other from the father. Thus, depending on the traits of their parents, children may have a variety of different genetic patterns. When the alleles are the same, we say there is a homozygous condition, and the alleles can be shown as RR or rr. There also is the possibility their alleles show the heterozygous condition, or Rr. Though there are three different possibilities for the genotypes (i.e., RR, Rr, or rr), there are only two different phenotypes (i.e., red or yellow). So, which genotype gives which phenotype? If the genotype is RR or Rr, the phenotype will be red, because the red is dominant to yellow. If

Table 14.1 Common Autosomal Disorders and Traits.

Dominant	**Recessive**

Amyotrophic lateral sclerosis (Lou Gehrig's disease)
  Degeneration of nerve cells
	Cretinism
	  Slowed physical and mental development, lack of thyroid tissue

Atopic allergic disease
  Hay fever
	Cystic fibrosis
	  Defect of exocrine glands producing mucous congestion of lungs, digestive disorders

Dimples
  Indentations in the cheeks and chins
	Phenylketonuria
	  Excess phenylalanine in the blood leading to mental deterioration

Earlobes
  Free are dominant over attached
	Sickle cell anaemia
	  Abnormal hemoglobin that crystallizes when exposed to low oxygen levels

Eye color Dark is dominant over blue (but with many other variables)	
Freckles Pigmented spots on the skin	
Gynecomastia Breast development in males	
Hair Curly is dominant over straight	
Hemochromatosis Iron overload, severe diabetes, fatigue	
Hypercholesterolemia Elevated blood cholesterol level	
Juvenile pernicious anemia Insufficient red blood cells	
Marfan's syndrome Body parts are long and thin, weakened connective tissue	
Ptosis Drooping eyelids	
Retinitis pigmentosa Deposits of pigment in the retina leading to blindness	

the genotype is rr, the phenotype will be yellow. Therefore, the dominant form covers up—or dominates—the recessive form.

Early geneticists would observe the traits of the organisms they were studying, and one technique they used to understand these observations was a **Punnett square**. The Punnett square also helps to predict the outcome of experiments with organisms to better interpret inheritance patterns. Looking at the traits does not always tell us what the alleles are. In other words, looking at the phenotype does not tell us what the genotype is. Therefore, geneticists perform crosses to help determine the original genetic make-up of the parents. A **cross** is a mating of two different parental stocks of plants or animals. After mating, the organisms produce offspring, and by observing the offspring's phenotype, the genotype of the parents often can be determined. Crosses can be simple, in which one trait is considered at a time (i.e., a **monohybrid cross**), or more complicated, in which two or more traits are considered at a time (i.e., a **dihybrid cross**).

Let's try an example of a monohybrid cross with our red and yellow pigments. We'll also use a Punnett square to show how it can be used. If we crossed a homozygous red individual with another of the same appearance, there would be little doubt regarding appearance of the offspring. Therefore, consider a cross between heterozygous red and homozygous yellow individuals, which means the red individuals are Rr and the yellow individuals are rr. To use a Punnett square, we must determine all the possible alleles that each individual can contribute to the cross. Regardless of which

Figure 14.1
Punnett Square showing the possible genotypes from a cross between a red individual and a yellow individual where the red trait is dominant.

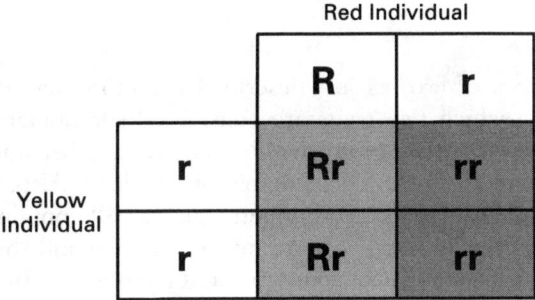

chromosome the yellow individual gets picked to be in the egg or the sperm, it will contain the r allele. This is not the case for the red individual, however, the red individual has two different alleles, R and r. Therefore, half the sperm or eggs will contain an R, and the other half will contain an r. The Punnett square will show the red individual's possible alleles on one side and the yellow individual's on the other (Figure 14.1).

After the possible alleles that can be contributed by each individual are recorded, the squares are drawn and then filled in with the combinations of alleles. Thus, in Figure 14.1, we see the possible genotypes of the offspring: Rr, Rr, rr, and rr. Note there is a duplication in the genotypes from the Punnett square. Half the offspring will have Rr, and the other half will have rr. Thus, there is both a genotypic and phenotypic ratio of 1:1. Half the offspring will be yellow, and half will be red.

Crosses and Punnett Squares

Objectives:
- To apply basic principles of inheritance to better understand them.
- To practice using Punnett squares to determine possible genotypes in a monohybrid cross.

Materials:
None.

Procedure:

1. Select a pair of traits that have a simple relationship to each other. Opposites make good choices, but other relationships also will work. As in the previous example, red and yellow are not opposites, but one occurs when the other is absent. Use traits such as tall and short, big and little, curly and straight.

2. Designate one item of the pair to be dominant and the other to be recessive. Remember to use the upper-case letter of the dominant allele and the lower-case letter for the recessive allele.

3. Perform all the possible crosses. Realize there are several possible crossessuch as:
    - A homozygous dominant individual with a homozygous dominant individual.
    - A heterozygous individual with a homozygous dominant individual.
    - A heterozygous individual with a heterozygous individual.
    - A heterozygous individual with a homozygous recessive individual.
    - A homozygous recessive individual with a homozygous recessive individual

4. Perform the crosses as described by first deciding what the possible alleles from each individual are. Construct the Punnett squares with enough squares to handle all the possible combinations. (Note that with a single trait, no more than four compartments are needed.)

5. Determine the genotypic ratio and the phenotypic ratio for each cross.

Dihybrid Crosses

The types of crosses just described are monohybrid crosses. Dihybrid crosses are those in which two traits are considered simultaneouly. Each time another trait is added for consideration, the Punnett square becomes larger, with more compartments, one for each new combination of alleles. Also, for each additional trait considered, the heading on the Punnett squares will contain another letter. Thus, if the red verses yellow example was combined with tall and short, in which tall was dominant over short, the possible genotypes and phenotypes for individuals to be crossed would be:

R = red   r = yellow   T = tall   t = short

RRTT	red, tall individual
RrTT	red, tall individual
RrTt	red, tall individual
rrTT	yellow, tall individual
rrTt	yellow, tall individual
rrtt	yellow, short individual

The Punnett square for crossing a homozygous red, tall individual with a homozygous yellow, short individual would be:

Red and Tall Individual

	(RT)
Yellow and Short Individual (rt)	RrTt

The list of possible crosses in a dihybrid cross will be much longer than that given for monohybrid cross. The cross listed here is one of the simplest, because both individuals are homozygous for both traits, with one being dominant and the other being recessive. Therefore, only one possible set of alleles can be passed on, thus making all the offspring heterozygous individuals. In the following exercise, you will perform several other crosses. You will decide the genotypes, the possible alleles that can be passed on, the phenotypes of the offspring, and the genotypic and phenotypic ratios of the offspring.

Dihybrid Crosses

Objectives:

- To understand how to extend the basic concepts of a monohybrid cross to deal with two or more alleles at the same time.

Materials:

None.

Procedure:

1. Complete the following crosses. Use Punnett squares to determine the genotypic and phenotypic outcome of each:

    - *Cross 1:* A homozygous red, homozygous tall individual with a heterozygous red, heterozygous tall individual.
    - *Cross 2:* A homozygous yellow, homozygous short individual with a heterozygous red, heterozygous tall individual.

2. Complete the following table with the results of the crosses:

	Possible Alleles	Genotypic Ratio	Phenotypic Ratio
Cross 1			
Cross 2			

QUESTION: One principle of inheritance is that each gene and trait will independently assort in the meiotic process. Many different genes are on each chromosome, however, meaning that not all genes can independently assort. These genes on the same chromosomes are called linkage groups. Explain what this means in relationship to the genes on the same chromosome as far as their pattern of inheritance.

Observations:

Conclusion:

Testcrosses and Pedigrees

As you know from the previous exercise, a homozygous dominant individual may look just like a heterozygous individual. The phenotype does not tell you enough to always know the genotype. In animal and plant genetics, a testcross can be performed to determine the genotype of an individual. A **testcross** is a cross between the unknown individual and a homozygous recessive individual. The outcome of the cross then can be used to determine the genotype of the unknown individual. If the individual with the dominant trait or phenotype and the individual with the recessive phenotype produce only offspring with the dominant phenotype, then the dominant-phenotype individual is homozygous. If the testcross produces some offspring that show the recessive phenotype, however, then the dominant-phenotype individual is heterozygous.

Not only do humans live too long to test cross, but there are many ethical and moral considerations as well. Therefore, testcrosses are not used in human genetics. Testing in which samples are taken from people and used to determine whether they have a particular allele has revolutionized genetic counseling, but this also has produced many medical, ethical, and moral dilemmas.

QUESTION: Knowing you or your spouse has a certain recessive gene that can cause a defect or genetic disease in your offspring would be quite disturbing. What are some of the ethical and moral dilemmas posed by this knowledge? Should we expand genetic testing, or should it be banned? What are the implications of genetic testing or screening for health care and health insurance?

Observations:

Conclusions:

Journal Note Do you know anyone with a genetic disease or disorder? Is someone in your family affected? (Remember that entries into your journal are private and will not be graded or read.)

Genetic testing is expensive and, in some cases, too expensive or still impossible. We do not know which genes control every aspect of development and functioning in humans. One technique used in humans is construction of a **pedigree,** which is a chart that shows the genetic connections or relationships between individuals of the same or closely related family groups. Figure 14.2 shows such a pedigree. There are rules that are used to make sure that pedigrees can be compared and studied, and Table 14.2 lists the commonly used symbols in pedigree construction. Thus, human genotypes may be determined by studying the trait of interest in the parents, siblings, and other close relatives of the affected individual. Medical records of patient histories and physical examinations are essential for this type of study.

Pedigree Analysis

Objective:

- To understand a basic tool of human genetic analysis.
- To understand pedigrees by interpreting an example.
- To construct a pedigree for a hypothetical family with a genetic disease.

Figure 14.2
Pedigree showing five generations of two families.

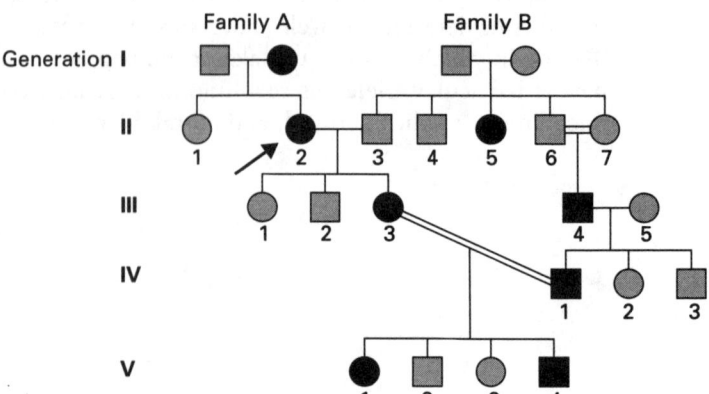

240   SECTION 14   Genetics

Materials:

None.

Procedure:

1. Using the symbol chart in Table 14.2, interpret the pedigree shown in Figure 14.2. Make sure you understand the relationships between individuals and their relatives. Generations are numbered as Roman numerals down the left side of the pedigree. The individuals in each generation are numbered with Arabic numbers so that they may be referred to reliably. You will probably not be able to determine the type of inheritance manifested in the pedigree based on the information in this manual and should refer to your text. The point of this exercise is to understand the relationships shown by the pedigree and how to construct them, not necessarily the genetic inheritance pattern.

2. Using proper symbols, construct a pedigree chart based on the following hypothetical family group:

    Generation I: This generation contains a carrier female and her brother. The female marries an unrelated male, and the brother dies at age 25.

    Generation II: The mating from generation I produces a carrier female, a normal female, an affected male, and a normal male. The normal female marries a normal male (i.e., mating 1). The carrier female marries a normal male (i.e., mating 2).

    Generation III: Mating 1 produces an affected male and a carrier female. Mating 2 produces two normal males.

    The affected male in generation II is the index case for this pedigree.

QUESTION: Genetic counseling often uses a combination of pedigrees and actual gene or DNA testing. Where would you obtain these services? Are these services provided anywhere near your home? Are they provided by the local county or state health department? Is this kind of testing covered by private insurance, Medicare, or Medicaid?

Observations:

Conclusions:

Journal Note What are some characteristics that are special in your family? In other words, do your parents, brothers, sisters or even close cousins have peculiar facial features that tell people you are related? Are there any twins, triplets, or larger multiple births in your family? (If there are multiple births in your family, are they fraternal or paternal?)

Table 14.2 Common Symbols in Pedigree Constructing

Symbol	Description
○	Normal female
□	Normal male
○—□	Single bar indicates mating
○-⊤-□ with ○ below	Out-of-wedlock mating and illegitimate child
○—//—□	Divorced
I: ○—□ ; II: ○ ○ □ (1, 2, 3)	Normal parents and normal offspring, two girls and a boy in birth order indicated by numbers; I and II indicate generations.
○═□	Double bar indicates a consanguineous mating
○△□ (twins)	Fraternal twins (i.e., not identical)
○△○ (twins)	Identical twins
○—□ with ○ below	Adopted daughter
◇	Child of unknown sex
● and ■ (with arrows)	Darkened square or circle indicates an affected individual; arrow indicates the index case or the beginning of the analysis.
⊙	Carrier
⌀ and ▱	Dead
◇ with P	Pregnancy

Appendix

Environmental Concerns

Most textbooks in Human Biology speak to the issues of the environment. Environmental concerns are prominent at all levels of government, on the agenda of national and international advocacy groups, and on many personal "lists of things to worry about." What do you know about your local environment? Is it threatened by all of the same dilemmas putting most cities at risk? Or is it facing a specific pressure from local industry, unique development projects, or environmental destruction?

This exercise is designed to guide you and your fellow investigators/consultants on your own environmental impact study. The lab class will be divided into three or four investigative teams depending on the size of the class. Each one will survey a different source of concern. The following is a list of possible issues:

- Air pollution
- Water supply/purification
- Waste water treatment
- Solid waste management, including recycling
- Special local issue

All communities are grappling with these, or related issues. What is happening in your college/university community or hometown regarding these areas?

Objectives:
- To develop interpersonal and team membership skills.
- To put environmental concerns into local, more comprehensible, terms.
- To become an expert on one area of environmental concern.

Materials:
- Record-keeping mechanism.
- Telephone directory and access to telephone.
- Internet access (optional).
- Meeting space.

Procedure:

Steps in the Environmental Impact Study Plan:

1. Schedule a planning meeting for your team. This should occur during part of a lab period so that all team members will be able to attend this first meeting. Subsequent meetings should be scheduled outside of class time so that the entire group, or various subcommittees can meet, discuss, and plan. Decide on the various subareas to be investigated within your area of concern, and divide the defined tasks among the members of the group. Meetings should be scheduled at least twice a week to help motivate the members of the group. Depending on the dynamics of the group and the rules set by your instructor, you may pick a team leader. Decide how, and in what form the information gathered will be presented. One member of the team may actually serve as a recorder for the group. This provides a second coordinating force besides the team leader. Exchange class and work schedules, addresses and phone numbers for ease of communication.

2. Suggested tasks/questions to be addressed:

    What agency (city, county or state) regulates your issue?

    Who are the contact people at the appropriate agency?

    What are the concerns of the community regarding this issue?

    Who bears the costs of this issue?

    How are the costs distributed between the citizens, local government, and the state?

    What is the capacity of the current facilities for this issue?

    What is the service level required for the community's needs?

    What is the projected life span of the facilities?

    What are the plans to deal with any projected shortfall in the ability of the current facilities to handle the needs? In the near future and in the distant future?

    Collect information regarding all of these areas and any additional ones developed in the early planning meetings.

3. Each member of the team should compile their information in whatever form agreed upon by the team members at the initial meeting. If one person is serving as the team recorder, make a copy of your information for the team leader and the recorder.

4. Each team will prepare a presentation on their specific topic. The presentation should cover the key background elements to put the issue into perspective; the major problems, if there are any, in dealing with this issue at the local, regional, state or national level, and the effects that this issue will have regarding the health and well-being of community members.

5. In your 10–15 minute presentation, you should be able to show that your team has thoroughly investigated the topic. If several people actually present information, there will be little doubt that all members of the group participated in the project.

6. An important part of any major project is evaluation. You should build an evaluation mechanism into your presentation/project. Ask the class to critique the presentation. To do this, formulate a short questionnaire that helps your audience evaluate the organization, thoroughness, and long-term benefits your environmental assessment has on your particular community concern.

7. Prepare and turn in a final report based on the guidelines provided by your instructor. Depending on the length of your required report, it may be difficult to include all of the materials/information that you collected. This means you will have to decide what to include, and what not to include. If there is vital information in the form of tables or charts that would be useful for future researchers, these can be included as appendices at the end of the report.

After completing this project, you and your team members will have a much better understanding of your particular issue. Depending on the size of your class and how well your classmates completed their reports, you will also learn about several other important issues related to topics vital to all of us.